Foundations of
Scientific Method:
The Nineteenth Century

Foundations of Scientific Method: The Nineteenth Century

Edited by
RONALD N. GIERE
and
RICHARD S. WESTFALL

INDIANA UNIVERSITY PRESS

Bloomington   *London*

Published in Canada by Fitzhenry & Whiteside Limited, Don Mills, Ontario
Library of Congress catalog card number 72–79910
ISBN: 0–253–32400–9
Manufactured in the United States of America

Contents

Preface

The coincidence of Indiana University's sesquicentennial and the tenth anniversary of the founding of the Department of History and Philosophy of Science provided the occasion, and part of the funding, for a joint celebration. Since we are irredeemably academic, the form of the celebration was pre-ordained—we organized a conference. The eleven essays in this volume, all original compositions for the occasion though not all delivered at the conference, are the fruits of the three-day birthday party attended by a substantial percentage of the historians and philosophers of science in the world today who are actively concerned with the topic of discussion.

That topic, which is also the title of the volume, was the Foundations of Scientific Method: The Nineteenth Century. A variety of considerations led to this choice. As a department combining the history and philosophy of science, we sought a topic that would at once reflect the dual nature of the department and actively engage both of its component disciplines. One possibility was a conference explicitly devoted to examining the mutual relations of history and philosophy of science. The issue is inevitably one of never ending concern to us; indeed it cannot be avoided by historians and philosophers of science regardless of their departmental affiliation. Nevertheless we decided against such a conference. Other conferences devoted to the question have been held, and there seemed to be more opportunity to contribute to understanding by building on the foundation already laid and finding a topic of significance to both on which the two fields could exhibit their mutual relations. The question of method was an obvious choice. Together with closely related problems, it is probably the central issue in contemporary philosophy of science. We (especially the philosophers among us) were convinced that historians interested in the development of scientific method could not fail to profit from the analytic tools philosophers of science command. Equally we (in this case, primarily the historians) were convinced that scientific methodology, like every other activity of man, is practiced in an historical context apart from which it cannot be fully comprehended. Moreover, historians of science generally agree that the development of scientific method is among the least well investigated aspects of the history of science. A well planned conference was likely to make a significant contribution in an area where the existing literature is neither extensive nor definitive. In a word, we found in scientific method a topic that filled all our requirements.

The choice of the nineteenth century followed almost inevitably. As we were aware—and as the range of questions pursued in this volume confirms—the development of scientific method is a subject so immense that a conference without some further limitations would have been hopelessly diffuse. In light of the existing state of the literature, we did not feel confident in imposing any but chronological limitations. A period earlier than the nineteenth century was not likely seriously to engage most philosophers of science; methodology in the twentieth century was apt to appear to most historians as primarily an exercise in contemporary philosophy of science. Moreover, the nineteenth century was a period of great intrinsic interest. The period witnessed an immense expansion of the scientific enterprise, now firmly established as the central intellectual feature of western civilization. Whereas the physical sciences had dominated the scientific revolution of the seventeenth century and furnished the instances of its methodological progress, other fields of science, such as biology and geology, now asserted their role and contributed their problems to methodology. The manifest success of natural science encouraged the growth of the social sciences, which now distinguished themselves from social philosophy, and which also offered new methodological problems. Beyond the problems raised by scientific practice, a number of philosophers with two centuries of scientific achievement on which to reflect pondered the issues of scientific method. Clearly the nineteenth century offered a considerable field for investigation—and one as yet hardly touched.

Beyond the limitation to scientific method in the nineteenth century, we made no effort to impose a common theme on the papers. As we have stated above, we were convinced that the present understanding of the question is such that any attempt to single out a central theme was unlikely to be fruitful. Indeed we question the possibility of ever defining the central theme of a development that was, as the papers in the volume illustrate, so multifarious. What common strand binds the methodological problems of geology to the reasoning employed by Maxwell in physics? What common strand binds the influence of *Naturphilosophie* on science to Marshall's work in economics? We make no pretense then that we are offering the definitive treatment of scientific method in the nineteenth century, or even that we are offering a history of it. We are bold enough to suggest that the papers in the volume do achieve the goals at which the conference aimed. On the one hand, they contribute significant new understanding of various aspects of a development that was complex beyond facile description. On the other hand, the very kernel of the contribution lies in the mutually supplementary insights of the philosophers and historians who brought their respective disciplines to bear on a problem of interest to both.

The conference was held in Bloomington on November 26–29, 1970. As we said above, a significant percentage of the philosophers and historians of science actively interested in the question gathered in Bloomington to reflect on it in conjunction with the speakers. By publishing the volume, we hope to extend the process of cross-fertilization.

R.N.G.
R.S.W.

Foundations of
Scientific Method:
The Nineteenth Century

1 Kant, Naturphilosophie[1] and Scientific Method

L. PEARCE WILLIAMS
Cornell University

"Was der Geist versprecht, leistet die Nature"—Oersted*

It seems appropriate to open a symposium on scientific method in the nineteenth century with a short discussion of the historical method adopted to attack the subject of this paper. The problem is this: Kant's philosophy, especially as revealed in his most read treatise, *The Critique of Pure Reason,* is not characterized by clarity and/or consistency. Kant's well known remark that he thought about the *Critique* for twelve years and wrote it in six months reveals at least part of the reason for difficulty. The fact that Kant "wrote" a good deal of the *Critique* by stitching together fragments of other treatises written years apart can account for the rest.[2] *The Critique of Pure Reason* is not a systematic, well thought out philosophical work but an agglomeration of philosophical reflections, often contradictory, loosely bound together by Kant's ideas on the nature of reason and reality. The result has been well summed up by one of Kant's commentators:

The profound influence which Kant has thus exercised upon succeeding thought must surely be reckoned a greater achievement than any that could have resulted from the constructing of a system so consistent and unified, that the alternative would lie only between its acceptance and its rejection.[3]

As the nineteenth century was to illustrate well, Kant could be used as a philosophical jumping off place by almost anyone and the philo-

* See note 43.

sophical results were often completely at odds with one another. It seemed folly to me, therefore, to attempt to uncover what Kant *really meant* and then examine the implications of this for scientific method. What Kant really meant or what, in Kant, is philosophically valid is a problem for philosophers. As a historian, I shall be concerned with what one group of thinkers in the early nineteenth century—the *Naturphilosophen*—thought Kant meant, for it was upon their interpretation of Kant that they built their system and their method. The result was a philosophy which Kant would undoubtedly have condemned, for it contained precisely those elements of speculation and unchecked Idealism which the *Critique of Pure Reason* was intended to hold up to critical scorn. Nevertheless, it was a "Kantian" edifice, for both the inspiration and the foundations came from Kant.[4] It was the *Critique of Pure Reason* that provided the *Naturphilosophen* with the basic justification for their philosophical approach. Kant's *Metaphysische Anfangsgründe der Naturwissenschaft*[5] offered a philosophical analysis of fundamental scientific concepts upon which the *Naturphilosophen* could erect their science of nature. As we shall see, both elements were influential in the creation of the method which was utilized by the *Naturphilosophen* and which stood in marked contrast to the methods employed by other scientists of the time.

One final methodological point deserves explicit mention. If, as I have hinted, every man could read into Kant almost what he wished, then what was it that bound the *Naturphilosophen* together into a coherent school? The cause may lie in Kant, for, in spite of his inconsistencies and contradictions, it is possible, as three generations of Kantian scholars have finally managed to show, to pull together a coherent Kantian philosophy.[6] But, in the absence of this century and a half of Kantian scholarship, it may be wiser to look for the unity of *Naturphilosophie* in the *Naturphilosophen* themselves. Were these men looking for the same answers in Kant and was this the reason that they coalesced into a "school"? Were there, in short, conditions in the European intellectual world of the late eighteenth century that can explain the emergence of *Naturphilosophie* and can give us the clues necessary to follow the *Naturphilosophen*'s idiosyncratic reading of Kant? The answer is yes, for the *Naturphilosophen* appear to be a singularly united group whose unity of viewpoint is inexplicable on any grounds other than common (and burning) concerns and common relief from these concerns in Kantian philosophy.

The question of who the *Naturphilosophen* were now demands an answer. If there is some kind of common ground shared by them all, it should be made clear here. First, let me enumerate those whom I shall be discussing. In alphabetical order, they are Samuel Taylor Coleridge,[7] Hans Christian Oersted,[8] Johann Ritter,[9] F.W. J. Schelling,[10] and Christian Samuel Weiss.[11] In the background, lurking so to speak in the shadows, but nevertheless discernible, I would like to place André-Marie Ampère[12] and, more hesitantly, Humphry Davy.[13] What, we must ask, does this group have in common that could lead them to adopt *Naturphilosophie* as a valid view of the world?

The first striking coincidence is birth dates. Coleridge was born in 1772; Oersted in 1777; Ritter in 1776; Schelling in 1775; Weiss in 1780; Ampère in 1775; and Davy in 1778. In spite of their national differences, they shared the experience of the French Revolution and they all reacted more or less similarly to it. There was the first youthful enthusiasm, followed by bitter disappointment. They were all severely affected by this experience, and they all felt it strongly. In varying degrees, they were all political conservatives in their adult years. The French Revolution also affected their religious sensibilities. The repulsion they felt for the excesses of the Terror was expressed in their horror of atheism, which they all saw as the cause of the political and social disintegration of Revolutionary France. Without God, there could be no social order, and no philosophy could claim their allegiance if God could not be made a central part of it. All, finally, were Romantics. This term is notoriously difficult to define but we can enumerate the characteristics shared by the *Naturphilosophen*. They were all highly sensitive men, seeking both beauty and truth in their philosophy. Most wrote poetry and expressed their emotions in verse. All had a deep sense of form and thought architectonically. The whole was always more important and more than the sum of the parts. All recognized the importance of and all felt the near ecstasy of creativity springing from the active mind. Spirit was as real to them as body. All underwent youthful crises and discovered Kant as the answer to their personal *angst*.

It is in this atmosphere that we must now turn to Kant. What I shall attempt is a Collingwoodian reconstruction. Given the characteristics listed above, what would an eager, young, passionate, devout philosophical aspirant find in Kant to soothe his soul, rest his mind, and guide his search for truth? From Kant's passages in the *Critique* on God, aesthetics, the nature of reality, and the nature of scientific theory could be drawn the material with which to construct a new approach to science and scientific method.

In the spiritual realm, Kant and the *Naturphilosophen* were of the same mind. The French *philosophes* and the English empiricists of the eighteenth century had together dealt what many considered to be fatal blows to the whole concept of Deity. The skepticism of Hume, the clockwork mechanism of the "Newtonian" scientists, and the sensationalist philosophy of Condillac and his followers had made it difficult, if not impossible, to preserve one's faith in a traditional God. At best, a First Cause could be posited but this offered little solace to men like Coleridge or Ampère (or Oersted or Davy) who took the existence of a personal God seriously. For them, Kant's *Critique* was like water in the desert. To be sure, Kant did not and could not prove that God existed, but he did the next best thing. He proved that no one could prove that God did not exist.

I do not at all share the opinion [Kant wrote] which certain excellent and thoughtful men . . . have so often been led to express, that we may hope sometime to discover conclusive demonstrations of the two cardinal propositions of our reason—that there is a God and that there is a future life. On the contrary, I am certain that this will never happen. For whence will reason obtain ground for such synthetic assertions, which do not relate to objects of experience and their inner possibility. But it is also apodeictically certain that there will never be anyone who will be able to assert the *opposite* with the least show [of proof], much less, dogmatically.[14]

Reason was incapable of both proof and disproof of God's existence. And, as Kant had written in the Preface to the Second Edition of the *Critique,* "I have therefore found it necessary to deny *knowledge,* in order to make room for faith."[15] Faith in God had important implications for science and for scientific method. It is this faith that led Kant to his famous doctrine of the philosophy of *als ob*. Although we cannot know that God exists,

we declare . . . that the things of the world must be viewed *as if* they received their existence from a highest intelligence. The idea is thus really only a heuristic, not an ostensive concept. It does not show us how an object is constituted, but how, under its guidance, we should *seek* to determine the constitution and connection of the objects of experience.[16]

This heuristic concept has two interesting and important corollaries. It first allows us to reestablish a place for speculative reason in scientific advance. Secondly, it firmly reintroduces teleology into the main body of science itself. Both these aspects of Kantian thought are worth more detailed consideration.

Speculation was the great philosophical bugaboo of both the eighteenth and nineteenth centuries. Speculative reason led to systems, and systems, inevitably, led to error. I feel certain that no creative scientist of any era would ever, privately, deny the importance of speculative thought, for this is the very bread upon which science feeds, but this was a far different thing from accepting speculation as a proper mode of scientific advance. Kant, however, could offer grounds for the justification of speculation. "This I do," he wrote, "by representing all connections *as if* they were ordinances of a supreme reason, of which our reason is but a faint copy."[17] Here is one of those cases where the *Naturphilosophen* read into Kant a good deal more than he intended. The main thrust of the *Critique of Pure Reason,* after all, was to show the *limits* of reason and to prevent the kinds of metaphysical leaps of fancy which Kant felt had tarnished philosophy for too long. But to the *Naturphilosophen* what counted was the reinstatement of the divinity of human reason. And, more importantly, the reinstatement of the human reason as capable of understanding the whole of creation, as a whole. We need contrast here only the role of human reason in Laplace's *Essay on Probabilities* and the rational faculty of an Idéologue with this view, to understand why the *Naturphilosophen* embraced Kant so warmly. Laplace indicated that only an infinite mind could comprehend the whole; the disciples of Condillac denied the very existence of a whole and insisted that all human reason could do was classify sensations and "tag" them with proper terms. But, for the *Naturphilosophen,* the idea that human reason was a reflection of Divine reason held out the intoxicating prospect of comprehension of the Divine idea and plan for this world. The metaphor, I insist, must be taken literally. A reflection, however dim, is, nevertheless, a copy of the real, and if one believed in the metaphor then one could, through speculative reason, participate in Divinity. It is this thought which served as both inspiration and guide to Schelling. His treatises on nature are exercises of the speculative reason in laying bare what, from considerations of Divinity, Schelling considered the nature of the world *must* be.[18] Schelling, too, founded the *Journal für Speculative Physik.* Another *Naturphilosoph,* Henrik Steffens, founded and edited the *Polemical Journal for the Furtherance of Speculative Physics.*[19] Both efforts were short-lived but they serve to illustrate the central role of speculative reason in *Naturphilosophie.*

It might seem as though I have here given valuable ammunition to those historians of science who insist that *Naturphilosophie* was both nonsense and inimical to the progress of science in the nineteenth century. What, after all, did Schelling or Steffens contribute? And is not a citation of their use of speculative reason all the proof that is needed that *Naturphilosophie* was a scientific dead end? Fortunately, it is possible to cite a creative scientist who made important contributions to science in defense of *Naturphilosophie* and in defense of speculative reason. He is, of course, Hans Christian Oersted.[20] In a discussion of scientific laws, Oersted stressed the consonance of human and Divine reason in a passage which, though lengthy, so well expresses the spirit of *Naturphilosophie* that it deserves reproduction here.

If we investigate these laws more closely we find that they harmonize so perfectly with Reason that we may assert with truth that the harmony of the laws of nature consists in their being adapted to the dictates of Reason, or rather, by the coincidence of the laws of Nature and the laws of Reason. The chain of natural laws, which in their activity constitute the essence of everything, may be viewed either as a natural thought, or more correctly as a natural idea; and since all natural laws together constitute but one unity, the whole world is the expression of an infinite all-comprehensive idea, which is one with an infinite Reason, living and acting in everything. In other words, the world is a revelation of the united power of Creation and Reason in the Godhead.

We now first comprehend how we can recognize nature through Reason, for Reason again recognizes herself in all things. But, on the other side, we can also conceive how our knowledge will never be more than a faint image of the great whole; for our Reason, although originally related to the infinite, is limited by the finite, and can only imperfectly disengage itself from it. No mortal has been permitted entirely to penetrate and comprehend the whole. Filled with devout awe, he must be conscious of the limits of his powers, and acknowledge that the feeble ray which he is permitted to behold, nevertheless raises him far above the dust. Yet we are not connected with the inward essence of nature by the clear sign of Reason alone. As in our taste for the Beautiful we receive a sense for the impression of the spirit in forms, and in the conscience a sense for the impressions of Reason in life, so we also receive a sense for the impressions of Reason in the operations of nature, by which we feel its proximity, and without a distinct view of the majesty of the whole. This anticipating consonance with Nature, guides reason in its inquiry, and is again awakened, strengthened, and purified by it; both are most intimately united, yet in such a manner, that the former is most dominant in life, the latter in science.[21]

Nor was this merely a philosophical point for Oersted. It was also a methodological guide. The "anticipating consonance with Nature" is what leads the human reason to discovery. Oersted explicitly opposed the methodology that argued that scientific laws were the result of abstraction from previously discovered facts.

The discovery of a natural law scarcely ever occurs by mere abstraction. It is a fortunate glance into nature by which the rule is discovered through which she acts. . . . It is with naturalists as with other artists; they think and act correctly in consequence of a fortunate suggestion, which they owe to a peculiar turn of mind, united to a closer and more intimate acquaintance with the matter. . . . The most beautiful discoveries in natural science have sprung from researches which were undertaken according to the demand of reason.[22]

Thus, to summarize, human reason is the dim reflection of Divine reason. Nature is constructed according to Divine reason. Hence, human reason can, unaided, comprehend Nature. But we must be aware that the reflection of Divine reason is dim in the human mind and therefore we must not trust naively what human reason reveals to us. Here is where Oersted parts company with Schelling and Steffens and introduces another element—experiment. We perceive what we think is a scientific law through reason but "we convince ourselves of its correctness by causing nature to act before our eyes, and to express her laws under the most different circumstances."[23] Sir Karl Popper would like that! Reason proposes and experiment disposes. And it all make scientific sense because there is a God who acts reasonably.

The second consequence of acting *as if* God existed is that we must also assume that there is (divine) purpose in God's creation. Kant spells out this teleological point explicitly.

This highest formal unity [he wrote], which rests solely on concepts of reason, is the *purposive* unity of things. The *speculative* interest of reason makes it necessary to regard all order in the world as if it had originated in the purpose of a supreme reason. Such a principle opens out to our reason, as applied in the field of experience, altogether new views as to how the things of the world may be connected according to teleological laws, and so enables it to arrive at their greatest systematic unity. The assumption of a supreme intelligence, as the one and only cause of the universe, though in the idea alone, can therefore always benefit reason and can never injure it. Thus if, in studying the shape of the earth (which is round, but somewhat flattened), of the mountains, seas, etc., we assume it to be the outcome of wise purposes on the part of an Author of the world, we are enabled to make in this way a number of discoveries. And provided we restrict ourselves to a merely regulative use of this principle, even error cannot do us any serious harm. For the worst that can happen would be that where we expected a teleological connection (*nexus finalis*), we find only a mechanical or physical connection (*nexus effectivus*). In such a case, we merely fail to find the additional unity; we do not destroy the unity upon which reason insists in its empirical employment.[24]

Teleology and opposition to it are two of the poles around which scientific controversy raged in the nineteenth century. The historian should be aware that the proponents of teleology were not fossils left over from the Middle Ages but were able to cite respectable philosophical authority for their views. What effect teleological beliefs had on scientific method in the physical sciences is difficult to discover because we know so little of the intimate beliefs of nineteenth-century scientists. Einstein's remark that he could not believe that God played dice with the cosmos might, under these circumstances, be taken as a somewhat more serious methodological statement than it usually is.

Teleology also reinforced another aspect of *Naturphilosophie* that is both important and easily documented. This is the deep belief in the essential unity of purpose and of structure in the universe. The cosmos, as a Divine creation, hangs together as the product of the Divine mind in a single organic whole. To know its parts, or to assume that it is enough to know its parts, is to miss its essential beauty, and beauty is as much a part of science as reason. Here again Kant's words fell on eager ears. Reason has two conflicting interests:

on the one hand interest in *extent* (universality) in respect of genera, and on the other hand in *content* (determinateness) in respect of the multiplicity of the species. . . . This twofold interest manifests itself also among students of nature in the diversity of their ways of thinking. Those who are more especially speculative are, we may almost say, hostile to the heterogeneity, and are always on the watch for the unity of the genus; those on the other hand, who are more especially empirical, are constantly endeavouring to differentiate nature in such manifold fashion as almost to extinguish the hope of ever being able to determine its appearances in accordance with universal principles.[25]

We can again turn to Oersted for echoes of this idea in action.

The first step in our investigation [of the spiritual in Nature] will be, to convince us that the laws of Nature, by which every individual thing is governed, not only forms [*sic*] a variety but a totality, a unity, and a whole.[26]

How this unity may affect our approach to Nature is discussed in another essay.

As a mental experiment, let us imagine that everything we know concerning the form of a sphere was still unknown, and that an artist endeavoured to discover a form that should appear alike on all sides, that should balance itself if placed upon a horizontal surface, should have a surface which would inclose a greater space than any other form of the same size: what an extraordinary depth and variety of thought it would require! But he, on the other hand, who starts from the principle of this form, viz. that of a space whose surface is everywhere equally distant from a centre, will find far more beautiful and remarkable properties from the necessary development of this idea, while a mere endeavour after this end, without a previous knowledge, would either never be successful or only by a circuitous means.[27]

It is this sense of the whole which is crucial. Without it, man can only enumerate the properties of individual things. And since the world is made up of an infinite number of individual things, such knowledge cannot ever advance to science which is knowledge of the whole. Again, the methodological implications are spelled out by Oersted.

To have seen a great number of natural phenomena is not to have an insight into nature. Experience only becomes instructive to us by a correct combination. To observe is to detect the actions of nature; but we shall not advance far in this path, unless we have a notion of its character. To make experiments is to lay questions before nature; but he alone can do that beneficially who knows what he should ask. Through the whole art of experience it is therefore necessary that, upon one side, the inquirer should constantly retain the whole in his view—for otherwise it is impossible to have a clear representation of the parts; on the other side, that he should regard nothing as beneath his attention, for it still belongs to the whole. . . . With this spirit and with this constant view of the whole, occupations which are frequently troublesome, and which enter into the smallest trifles, lose their insignificance to him; he elevates them to himself, and does not allow himself to be drawn down by them. He does not content himself with a single one-sided experience. He seeks everywhere to combine it with others, to deduce the one from the other, and to arrange all in such a manner, that the whole course of observations or experiments represent one natural law. . . . It is only by giving the observations and experiments which are made such a *connection, such an extension and variety,* that his labour can procure him knowledge, and become more than an imperfect account of an isolated phenomenon.[28]

Unlike the Baconians, what the *Naturphilosophen* were after was, in Goethe's words, "Das Was bedenke, mehr bedenke Wie."[29] The form is at least as important as the substance; theory clearly is superior to fact and theory was determined by its relation to the whole.

It might here be objected that Oersted and the *Naturphilosophen* were creating a methodological dichotomy between those who merely collected facts and those who could interpret them by a correct combination. Is not the correct combination inherent in the facts themselves? There is, after all, only one Nature and if enough facts are collected surely their "correct combination" will automatically emerge from the collected mass. Thus Oersted's distinction would appear to be merely a matter of convenience, not an important point of method. This is what a good Baconian or inductivist would say, but it is wrong. It would be true if facts were "things in themselves" and therefore the actual existences which we could know directly. But phenomenal facts, as Kan insisted over and over again, are *not* "things in themselves" but mere appearances. As the *Naturphilosophen* read him, the connection of these appearances depends upon the human mind, not upon Nature. Hence it is not only possible, but probable, that men will get the connections wrong if they naively assume that they are dealing with reality instead of their own psychological states. Kant's position here deserves citation.

That nature should direct itself according to our subjective ground of apperception, and should indeed depend upon it in respect of its conformity to law, sounds very strange and absurd. But when we consider that this nature is not a thing in itself but is merely an aggregate of appearances, so many representations of the mind, we shall not be surprised that we can discover it only in the radical faculty of all our knowledge, namely in transcendental apperception, in that unity on account of which alone it can be entitled object of all possible experience, that is, nature.[30]

And again:

Thus the order and regularity in the appearances, which we entitle *nature,* we ourselves introduce. We could never find them in appearances, had not we ourselves, or the nature of our mind, originally set them there. For this unity of nature has to be a necessary one, that is, has to be an *a priori* certain unity of the conception of appearances; and such synthetic unity could not be established *a priori* if there were not subjective grounds of such unity contained *a priori* in the original cognitive powers of our mind, and if these subjective conditions, inasmuch as they are the grounds of the possibility of knowing any object whatsoever in experience, were not at the same time objectively valid.[31]

It is possible to interpret these and similar passages in various ways, but all interpretations must grant specific and important original powers to the human mind in the formulation of scientific law. Schelling built upon this subjective foundation a towering edifice of speculative physics almost completely divorced from experience; Oersted recognized the danger of free speculation and insisted upon the correct use of experiment to hold speculation in check. There is also the possibility of other procedures. I shall here note only one, that devised by Ampère. Since we cannot know "things in themselves" and since the human mind drives itself to unity and scientific law, Ampère argued that it was justifiable to speculate freely within rather clearly specified limits. These limits are set by phenomena. What we can—indeed, must—do is to speculate on the nature of "things in themselves" or noumena and from these speculations reconstruct the phenomenal world.[32] Thus Ampère felt it perfectly legitimate to assume his electrodynamic molecular model with its material components and currents of both positive and negative electricity because from it he could deduce the magnetic phenomena of the sensible world. When his critics asked him for evidence for the existence of his noumenal model, he could not, of course, offer any. By definition, no such evidence can ever be cited. He could only offer his model as a possible noumenal *cause* of easily observed phenomena. Speculation here served to create a noumenal model—a theoretical entity—which introduced both unity and precision into the theory of magnetism. Unity and precision were justification enough.

Naturphilosophie was not content to leave the matter on this level. Granted that the human mind sought unity and imposed it upon appearances, granted, too, that speculation or imagination could lead to such unity, was this enough? For Schelling and those who really believed in the reliability of the speculative reason, the answer was yes. For Ampère, provided that the unity was capable of mathematization and was consonant with phenomena, the answer also was yes. But for Oersted, Ritter, and Weiss, the answer was no. It is perhaps significant that these three *Naturphilosophen* were chemists and involved with the relations of matter on the molecular level. Ampère could assume a molecule in order to deduce phenomena on the macroscopic level, but the nature and activity of the molecule in its own microscopic realm was what Oersted and his friends were after. They could not be satisfied with a model made up arbitrarily; they required a much closer fit, both physically and philosophically, between their assumptions and the nature of the phenomena they were attempting to explain. Once again, it was Kant who provided them with the necessary philosophical basis for their theories. After demonstrating in the second antinomy of pure reason that it is impossible to decide between an atomic and plenist view of the world, Kant nevertheless suggested a way out. In the *Critique,* the way was only dimly lighted but it was developed further in the *Metaphysische Anfangsgründe der Naturwissenschaft.*[33] In the *Critique,* Kant pointed out that

We are acquainted with substance in space only through forces which are active in this and that space, either bringing other objects to it (attraction), or preventing them penetrating into it (repulsion and impenetrability). We are not acquainted with any other properties constituting the concept of the substance which appears in space and which we call matter.[34]

The *Metaphysische Anfangsgründe* expanded upon this epistemological point and laid the foundation for the "dynamic" physics of the early nineteenth century. It was this physics of forces that served to pull all the disparate elements we have already discussed together into a unified system.[35] And it was this unity that made *Naturphilosophie* so attractive to a generation of scientists. It was also from this unity that the most important methodological innovation of *Naturphilosophie* was drawn. Oersted must again serve as our guide, for it was he who best articulated the position of *Naturphilosophie.*

Kant's original attractive and repulsive forces were adopted and transformed by Schelling. What Schelling added was the substitution of development for equilibrium between conflicting forces. Thus when equal and opposing forces met the result for Schelling was not stable equilibrium, but development into a higher conflict. It was in Schelling that the famous triad of thesis, antithesis, and synthesis first appeared in the guise of dynamic, conflicting forces. The world is thus the scene of ceaseless activity and dynamic progression. It is not and cannot be the dead, mechanical system of the materialists. Everything is in constant activity[36] and this activity is sustained by God. "He incessantly creates the entire infinite manifold existence, and this lives in him." God literally manifests himself in forces, and these forces can and do include spirit as well as matter.[37] To return to our earlier discussion of God and Truth, we can now add the presence of God *in* the world as a further buttress for our belief in its rationality, "for all existing objects are active forces of nature, which represent to us a unity of thought."[38]

From the conflict of forces also comes beauty. "Symmetry," Oersted tells us, "is one of the most comprehensive forms of beauty to the inhabitants of the earth, but it is founded on one of the principal features of thought, the unity of opposites."[39]

Truth and Beauty derive from God's ceaseless activity manifested through conflicting and opposing forces. So, too, do physics and chemistry. In the only book he ever wrote, translated into French under the title, *Recherches sur l'identité des forces chimiques et electriques,*[40] Oersted showed how it was possible to create a dynamic chemistry. More importantly, he there predicted the discovery that was to give him scientific immortality—the transformation of electrical into magnetic force. His discussion of the possibilities of this transformation revealed a new method of discovery that was to prove extremely fertile. It followed rather simply from the doctrine of conflicting forces: conflicting forces, under certain conditions, as Schelling had insisted, transform themselves into other forces. Oersted had used this principle to explain oxidation, acidification, and neutralization in chemistry and now extended it to the "pure" forces of electricity and magnetism. His chemical researches had convinced him of the unity of force; all that remained to be done was to determine the *conditions* under which the transformations of force take place.[41] Oersted's conception was literally protean, for he suspected that forces, like Proteus, changed only when confined and tortured. Thus it was that he was led to his famous experiment in the winter of 1819–1820 in which he passed an electric current through a thin wire of high resistance, convinced that this "confinement" of the electric force would force it to change into magnetism.[42] The experiment succeeded, and it set off a search for the transformation of forces which marked the years 1820–1860 as ones of dramatic discoveries. The new method was astonishingly successful. One need only observe it in the hands of a master such as Michael Faraday to realize its fruitfulness. I can particularly refer you to Faraday's search for the "magnetization" of light as an example of what this method entailed. Experiment after experiment was carried out. Note what these experiments did *not* do. They did not confirm an earlier speculation—most failed to produce the expected result. They did not refute a previous hypothesis—Faraday never interpreted failure to detect an effect as proof that his hypothesis was incorrect. What failure meant was that the correct conditions for the effect to be produced had not yet been set, and Faraday set out upon an almost Baconian "collection of instances" in the hope of discovering the proper experimental set-up. No number of unsuccessful experiments could prove anything except that the right conditions for the production of the expected effect had not yet been achieved. I do not know what to call this use of experiment, but it is worth noting that it is not defined by any existing philosophy of science. It is not induction, at least as that term is generally understood. Nor is it refutation. It gets one no nearer, either, to problem solving. Yet it was used to advantage by those who sought the conversion of forces. It might, therefore, repay further philosophical and historical study.

Here, properly, we should leave the subject. I think I have shown that Kant and *Naturphilosophie* did produce a scientific method peculiar to *Naturphilosophie* and of obvious importance to the historian of nineteenth-century science. But there was a final effect that deserves mention. Kant, and more particularly the *Naturphilosophen,* attempted to substitute a new cosmic metaphor. The world of the eighteenth-century *philosophe* was a machine; the *Naturphilosophen* insisted it was an organism. Its laws were laws of development; its basic theoretical paradigm was field theory in which the connections between parts were as important as the parts themselves. Organisms live because they are informed by Spirit, and the *Weltseele* was the ultimate substratum of physical reality. Only spirit can understand spirit; science, then, is spiritual in its essence. Unlike Faust, the scientific Romantic need not sell his soul to comprehend the Universe; he need only understand it.

Oersted, more than once, expressed this aspect of *Naturphilosophie* by a quotation he attributed to Schiller. It may well serve as the motto of this paper:

"Was der Geist versprecht, leistet die Natur."[43]

NOTES

1. I shall deal here only with *Naturphilosophie* as it applies to physics. The general philosophical framework of *Naturphilosophie,* as here presented, however, applies equally well to biology.

2. For a discussion of the assembling of the *Critique of Pure Reason,* see Norman Kemp Smith, *A Commentary to Kant's "Critique of Pure Reason,"* 2nd ed. (New York, 1962), p. xix ff. It is, perhaps, not out of place here to mention how I have read Kant. I have used the Insel-Verlag edition of Kant's *Werke,* edited by Wilhelm Weischedel, 6 vols. (Wiesbaden, 1960), simply because I own it and it was, therefore, the most convenient edition at hand. Since this paper contains no knotty problems of textual interpretation, use of this edition seemed justified. The standard and complete edition of Kant's *Werke* is that published by the *Königlich Preussischen Akademie der Wissenschaften* which contains excellent and full critical apparatus. I have also unashamedly used the "pony" provided by Norman Kemp Smith in his translation, *Immanuel Kant's Critique of Pure Reason* (London, 1958). Both Professor Smith's German and his technical knowledge of philosophy are better than mine and I found him a great help with difficult passages. All translations of Kant that follow are taken from Smith. I have checked them against the original and am unable to improve upon them. I have, in the references to these citations, included the pagination for the First (A) and the Second (B) editions of the *Critique.*

3. Smith, *Commentary,* p. 317. This is important and deserves emphasis. Kant meant different things to different people and the unsystematic nature of the *Critique of Pure Reason* allowed a wide variety of interpretations of Kant's philosophical position.

4. This seems to be a difficult point for philosophers to grasp, for the almost universal reaction of the philosophers of science at the conference to this paper was that I had misread Kant. Let me state explicitly that I have misread Kant—deliberately. I have, however, tried to misread Kant in the same way as the *Naturphilosophen* misread him. The problem is an historical, not a philosophical, one. Let me state it in its starkest outline. Kant's philosophy, as presented in the *Critique of Pure Reason* is obscure, often contradictory, and capable of being understood in a number of different, often contradictory ways. The group of men I shall discuss found in the *Critique* a philosophical justification for their own views, which views are unquestionably "un-Kantian" according to later interpreters of Kant. The questions I wish to answer in this paper are: what were the historical circumstances that led this group to read Kant in the way they did? What was the *Weltanschauung* this group

shared and which they felt their reading of Kant supported? Finally, what effect did their adoption of "Kantian" views have upon the way they approached the natural world?

5. Insel-Verlag edition, vol. 5. There is no trustworthy translation of this treatise.

6. It would be impertinent of me here to attempt a bibliographical overview of Kantian scholarship. I think it appropriate, however, to mention those works on Kant which I have consulted and found useful. Hans Vaihinger, *Commentar zu Kants Kritik der reinen Vernunft,* 2 vols. (Stuttgart, 1881–92); Ernst Cassirer, *Kants Leben und Lehre* (Berlin, 1921); Erich Adickes, *Immanuel Kants Kritik der reinen Vernunft* (Berlin, 1889), and *Kants Opus Postumum (Kant-Studien, Ergänzungshefte* 50, Berlin, 1920).

7. The most valuable sources for the study of Coleridge's encounter with Kant are his own *Biographie Literaria.* I have used the edition by J. Showcross, 2 vols. (Oxford, 1958), see especially 1: 88–105. The details of Coleridge's debt to Kant are in S. T. Coleridge, *Notebooks,* ed. Kathleen Coburn, 2 vols. (New York 1957–61), 1, and in Samuel Taylor Coleridge, *The Philosophical Lectures* (1818–19), ed. Kathleen 1956–59), 1. For Coleridge's later philosophical views see Samuel Taylor Coleridge, *The Philosophical Lectures* (1818–19), ed. Kathleen Coburn (London, 1949).

8. There is no adequate biography of Oersted. See the biographical introduction by Kirstine Meyer to H. C. Orsted, *Scientific Papers,* 3 vols. (Copenhagen, 1920). Oersted's debt to Kant is explicitly acknowledged at the end of his doctoral dissertation, "Grundtraekkene af Naturmetaphysiken tildeels efter en nye Plan," (Copenhagen, 1799), (*Scientific Papers,* 1: 35 ff.).

9. There is a useful sketch of Ritter's life by Armin Hermann which serves as the introduction to his volume on Ritter in *Ostwald's Klassiker, Die Begründung der Elektrochemie* (Frankfurt am Main, 1968). For Ritter's philosophical views, see J. Ritter, *Fragment aus dem Nachlasse eines jungen Physikers* (Heidelberg, 1810). For the incorporation of these views in his scientific work, see J. W. Ritter, *Das electrische System der Körper* (Leipzig, 1805). It should be noted that Ritter's works are extremely rare. The only complete set I know of is in the Ronalds Library of the Institution of Electrical Engineers, London.

10. Other than the philosophical works mentioned below (see note 18), in which Schelling makes his debt to Kant clear, see his *Zur Geschichte der Neueren Philosophie* (Stuttgart, n.d.), pp. 66 ff.

11. Weiss's indebtedness to Kant is made clear in his correspondence with Oersted. See *Correspondance de H. C. Orsted, publiée par M. C. Harding,* 2 vols. (Copenhagen, 1920), 1: 255 ff. It is worth noting that

Ritter dedicated his *Elektrische System* to Weiss and that Oersted was heavily influenced by Ritter.

12. See my article on Ampère in the *Dictionary of Scientific Biography* (New York, 1970). For Ampère's philosophy, see also André-Marie Ampère, *Essai sur la philosophie des sciences,* 2 vols. (Paris, 1838–43); and J. Barthélemy Saint-Hilaire, *Philosophie des deux Ampère* (Paris, 1866).

13. I have attempted to connect Davy with *Naturphilosophie* in my *Michael Faraday, A Biography* (London and New York, 1965). I have there argued that Coleridge was the medium through which Kantian ideas were conveyed to Davy, for Davy and Coleridge were good friends and often discussed metaphysics together. See Coleridge's letter to Davy in Coleridge, *Collected Letters.*

14. *Immanuel Kant's Critique of Pure Reason,* trans. Norman Kemp Smith (London, 1958), p. 595. A 741. B 769.

15. Ibid., p. 29. B xxx.

16. Ibid., p. 550. A 670–1. B 698–9.

17. Ibid., p. 555. A 678. B 706.

18. See J. F. Schelling, "Von der Weltseele, eine Hypothese der höheren Physik zur Erklärung des allgemeinen Organismus"; "[Einleitung zu den] Ideen zu einer Philosophie der Natur"; "Erster Entwurf eines Systems der Naturphilosophie"; "Einleitung zu dem Entwurf eines Systems der Naturphilosophie oder über den Begriff der spekulativen Physik und die innere Organisation eines Systems dieser Wissenschaft" in *Schelling's Werke,* ed. Manfred Schröter, 9 vols. (München, 1927), 1: 413, 653; 2: 1, 269.

19. For Steffens, who knew all the *Naturphilosophen* and wrote amiably about them, see Henrik Steffens, *Was ich erlebte,* 10 vols. (1840–44).

20. For Oersted and the influence of *Naturphilosophie* upon him, see the well known articles by Robert Stauffer, *Isis* (1953), 44: 307 and (1957), 48: 33. See also the recent article by Gerhard Henneman, "Der Dänische Physiker Hans Christian Oersted und die Naturphilosophie der Romantik," *Philosophia Naturalis* (1967), 10: 112; and Hennemann, *Naturphilosophie im 19. Jahrhundert* (Freiburg/München, 1959).

21. H. C. Oersted, "On the Spirit and Study of Universal Natural Philosophy," in *The Soul in Nature,* trans. Leonora and Joanna B. Horner (London, 1852; reprint, London, 1966), p. 450.

22. "Natural Science in its Relation to Different Periods of the World, and to the Philosophy Prevalent in Them," ibid., p. 285.

23. Ibid.

24. Kant, *Critique,* p. 560. A 686. B 714.

25. Ibid., p. 540. A 655. B 683.

26. "The Spiritual in the Material. A Conversation," *Soul in Nature,* p. 20.
27. "Superstition and Infidelity in their Relation to Natural Science," ibid., p. 88.
28. "On the Spirit and Study of Universal Natural Philosophy," ibid., p. 457.
29. *Faust,* (Tübingen, 1808), Part II, line 6992.
30. Kant, *Critique,* p. 140. A 114.
31. Ibid., p. 147. A 125. Here again the modern interpreter of Kant would argue, probably correctly, that this was not what Kant meant. What is historically important is the fact that the *Naturphilosophen* did think it was what he meant and constructed their philosophies accordingly.
32. See my article on Ampère in the *DSB* for a somewhat lengthier discussion of this point.
33. This treatise was written at the same time as Kant was preparing the second edition of the *Critique.* He was in full possession of his powers and his views deserve both our attention and our respect. It is a curious fact that few of Kant's later commentators take this treatise seriously. It was the fundamental tract for *Naturphilosophie,* for it was upon the Kantian doctrine of opposing forces that *Naturphilosophie* was to be erected.
34. Kant, *Critique,* p. 279. A 265. B 321.
35. It is true that dynamic physics itself would split into two schools in the nineteenth century. There were those who followed the Kant of the *Met. Anf.* and insisted that forces emanated from points and who could reconcile their views with those of more conventional atomists. There were others who failed to see the necessity of such centers of force and who were, therefore, hostile to atomism in any guise. This difference may be explored in the correspondence between Weiss and Oersted, *Correspondance de H. C. Örsted,* 1: 257 ff. It remains true that both schools drew their inspiration from Kant and both agreed upon the fundamental reality of forces.
36. See Oersted, *Soul in Nature,* pp. 3 ff.
37. Ibid., p. 4.
38. Ibid., p. 22.
39. "All Existence a Dominion of Reason," *Soul in Nature,* p. 111.
40. Paris, 1813.
41. I am grateful to Dr. Joseph Agassi of Boston University for many enlightening and exciting conversations on this subject. It is he who has first recognized this search for conditions as an essential part of the method of "dynamic" science.
42. See Oersted's account in H. C. Ørsted, *Scientific Papers,* 2: 351 ff.

43. Oersted, *Soul in Nature,* p. 13. It seems fitting to end this paper by a misreading. Schiller, so far as I have been able to discover, never wrote this line. What he did write was:

> Mit dem Genius steht die Natur in ewigem Bunde,
> Was der Eine verspricht, Leistet die andre gewiss,

(I am indebted to Miss Elizabeth B. Wilkinson, the great Schiller scholar for this attribution.) The line appears at the end of his poem, "Columbus," vol. 1 of the Nationalausgabe of Schillers *Werke,* p. 239. In the German edition of Oersted's *Der Geist in der Natur,* 2 vols., trans. Prof. Dr. K. L. Kannegiesser (Leipzig, 1854) p. 65, the line reads: "Was der Geist verspricht, das hölt die Natur."

The alteration of the original, in Oersted's mind, illustrates the force with which he held to the spiritual nature of physical reality.

2 Leading Principles and Induction: The Methodology of Matthias Schleiden

GERD BUCHDAHL
University of Cambridge

I

It is appropriate that at a symposium devoted to a discussion of methodological ideas in the nineteenth century, Matthias Schleiden (1804–1881) should receive honorable mention, for Schleiden presents the relatively rare phenomenon of a creative scientist who developed his ideas within the context of a consciously organized methodological framework, explicitly formulated to that end, and of a complexity rarely matched until recent times. To the historian of biological science, Schleiden, together with Theodor Schwann, is known as one of the key figures in the origins of modern cell theory. As Schleiden puts it in his celebrated memoir, "Contributions to Phytogenesis," every plant and animal organism is "an aggregate of fully individualized, independent, separate beings," i.e., cells.[1] Furthermore, as foreshadowed in the same memoir,[2] and generalized subsequently by Schwann, who here came under Schleiden's immediate influence, the different animal and plant forms "originate only from cells,"[3] which are the basic organ of development; "all cells grow according to the same laws," and the "cause of growth" lies in the individual cells and not in the organism as such.[4] This theory involved the rejection of a vitalist–teleological approach, in accordance with Schleiden's pronouncement that "the apparent gap between inorganic and organic form is not unbridgable."[5] As Schwann formulated the same position,

The fundamental powers of organised bodies agree essentially with those of inorganic nature, . . . they work altogether blindly according to laws of necessity and irrespective of any purpose, . . . they are powers which are as much established with the existence of matter as are the physical powers.[6]

Clearly, behind such theories there must lie a number of methodological assumptions and, indeed, philosophical positions. The interest of Schleiden's work consists in the clear and explicit manner in which these views were stated in his major writings and in the intellectual trends to which they point. Hence they may help the historian of ideas to trace interrelations between some of the contemporary movements of general philosophical thought and the development of the redolent scientific notions.[7]

Since Schleiden was a scientist of considerable stature, it is of particular interest to start by noting the interplay between his general assumptions and his own scientific activity. For example, general presuppositions, such as the idea of continuous transition between the inorganic and organic realms, frequently dominate the record of his observations or, perhaps we should say, what—with our knowledge of hindsight—Schleiden *seemed* to observe. Thus he has a complicated descriptive report about one form of cell growth, according to which the cell nucleus—he was responsible for this term and for making it central to his theory of growth—crystallizes out of the nourishing solution, with the cell forming round the nucleus and enclosing part of the solution.[8] This account, vaguely indicated in "Contributions," appeared more explicitly in Schleiden's major work, *The Principles of Scientific Botany*.[9] But when we look more closely there, we find that the core of the "observational report" is heavily indebted to analogical reasoning. As Schleiden states, referring himself to Schwann's *Researches,* there is an "interesting comparison between the formation of crystals and of the cell."[10] The statement echoes Schwann's position, according to which "the process of crystallization in inorganic nature . . . is the nearest analogue to the formation of cells."[11] The idea continued to cast its shadow, for even when the "correct" theory of growth by cell division could no longer be resisted, Schleiden still let his own account stand side by side with that of his critics, probably because its plausibility was supported by the organic–inorganic transition concept. One of Schleiden's early biographers concluded that his theory of cell formation was a kind of "deduction" from the analogy with crystallization.[12]

Of course, it would go too far to suggest that it was only a piece of analogical reasoning that led Schleiden to his mistaken views. He was also misled by the ordinary course of development in certain special plant cells, and by the misfortune (as Hughes points out in his *History of Cytology*) of "taking as his typical example of cell formation in a developing plant the events within the embryo-sac after fertilisation."[13]

In all this, what is of interest to us, however, is the fact that Schleiden, who makes the place of general presuppositions central in science, is not yet always clear on the way in which they may insinuate themselves into so-called "observations." Although he explicitly claimed to have represented accurately what "I have observed in most of the plants which I have investigated,"[14] his claim should be compared with the verdict of one of the most eminent of nineteenth-century historians of botany, Julius Sachs, whose classical *History of Botany* includes a detailed and usually sympathetic treatment of Schleiden's researches. Of Schleiden's "description" of cell growth in "Contributions," Sachs says:

He who is acquainted with the modern view of the processes of free cell-formation founded on numerous and careful investigations of later times will scarcely discover in the above account of Schleiden's theory a single correct observation.[15]

Of course, such an inductivist judgment is rather harsh, for we are now more keenly aware of the theory-ladenness of observation; the judgment seems especially harsh when we consider the crude state of microscope technology during the 1840s. Moreover, it is interesting to note that when he wishes to defend another of the great botanical figures of the same period, the redoubtable Hugo von Mohl, himself a severe critic of Schleiden, Sachs has a far more adequate view of the logic of the situation which he might well have applied to Schleiden also. It is still worth citing today:

These and some other errors on the part of a gifted and truly inductive enquirer are instructive, since they show that observation without any

ground-work of theory is psychologically impossible; it is a delusion to suppose that an observer can take the phenomena into himself as photographic paper takes the picture; sense-perception encounters views already formed by the observer, preconceived opinions with which the perception involuntarily associates itself. The only means of escaping errors thus produced lies in having a distinct consciousness of these presuppositions, testing their logical applicability and distinctly defining them.[16]

This was written, not in 1965, but in 1875! The only fault today's reader might find with this passage is its still insufficient recognition of the necessity of presuppositions that cannot be eliminated, to the extent that they infiltrate into the so-called "observational core." Nor is Schleiden much clearer on this aspect, even when speaking *ex cathedra* as a methodologist. Certainly, in those passages of his methodological pronouncements where he discusses methods of observation, we find continual insistence on careful and accurate observation, without any discussion of the problem of the theoretical ingredient. When holding forth in this vein, he certainly does not appreciate the weight of the inorganic analogy on his account of cell formation, and he has little inkling (and who would blame him) of the importance of the question of the physico–chemical stability of the smallest units of cell growth.

Still it is not so much his actual methodology that is at fault, but rather its insufficient application. Since he is a thoroughgoing Kantian, his basic approach to the theory of knowledge holds that all empirical judgments are the resultant of three components of knowledge–spatio–temporal form, leading to a mathematical formulation of knowledge, logical concept, and finally, sensory perception. It is certainly of interest to note that in this connection he draws the sort of conclusion which has become a commonplace in recent philosophy of science, and which he states by saying that "almost every proposition which we express in ordinary life is already an incomplete theoretical whole."[17]

II

Let us now take a closer look at the principles that underlie Schleiden's method. If occasionally he speaks with a somewhat inductivist voice, not appreciating the ingression of theory into "observation," particularly since he is working in an altogether novel field, nevertheless his influence on botany was pronounced, precisely because his views were expressed against a far wider intellectual background than was usual for the scientists, let alone botanists of his time. As Cassirer remarks in his *Problem of Knowledge,* "Schleiden was striving for nothing less than the setting up of a theory and critique of science."[18] He quotes the judgment of Sachs, that "his great merit as a botanist is due, not to what he did as an original investigator, but to the impulse he gave to investigation" as such.[19] It is noteworthy that Schleiden's views on the details of cell theory and cell growth, as well as fertilization, though usually "wrong-headed" (as we should judge them now), invariably set up a chain reaction, leading to large-scale research as a critical response to his writings, and eventually resulting in more adequate accounts. As Sachs tells us, all "these observers [i.e. Nägeli, von Mohl, Unger, et al.] were chiefly concerned to test the correctness and general applicability of Schleiden's theory."[20] I think that this is often due, not only to Schleiden's clear, and frequently critical and aggressive style, but also to the larger vistas and points of view opened up by his writings.

This becomes clear already from an inspection of his first major paper, the "Contributions." Sachs does not appreciate the importance of the general framework. All he has to say, rather sarcastically, is that "the work begins with some remarks on the general and fundamental laws of human reason."[21] As philosophical critics rather than inductive historians, however, our ears at once prick up. We are not prepared any longer to dismiss such notions as extraneous to the enterprise of the scientist. Alas, few of those who have written about Schleiden up to the present have shown much appreciation of the relevance of the philosophy behind the methodology, let alone behind the science. Part of the reason must be that Kant's metaphysics of nature, which Schleiden seeks here to resuscitate, had for the most part been transformed into the *Naturphilosophie* of Fichte, Schelling, and Hegel, and has met with little sympathy from those who were vehemently opposed to that philosophy. On the other hand, the untransformed and genuine core of Kant's philosophy of science had been lost in the puzzles of transcendental scholarships; and with the exception of a few, rather summary remarks from Cassirer, little has been done to utilize it for an understanding of its relevance to Schleiden's own thinking.

Let us ask then, first, what is this "general fundamental law of human reason," mentioned by Schleiden in the opening paragraphs of his paper? It is, Schleiden writes, "its undeviating tendency to unity in its acquisition of knowledge" which reason has always "evinced in the department which treats of organized bodies as fully as in all other branches of science."[22] How can this "unification" be brought to bear on the subject of plant life? By attending to the "idea of the individual"—which leads Schleiden to contend that the representation of individuals, all alike (note: a quasi-atomistic approach), can only be realized at the level of the cell. And generalizing by a surprising leap from the case of *Algae* and *Fungi,* he concludes that it is

easy to perceive [sic] that the vital process of the individual cells must form the very first, absolutely indispensable fundamental basis, both as regards vegetable physiology and comparative physiology in general; and, therefore, in the very first instance, this question especially presents itself: *how does this peculiar little organism, the cell, originate?*[23]

No mean "unification," this! What we need to appreciate is that this invocation of unity as a "fundamental law of human reason" was not an occasional *ad hoc* introductory gloss, but a by-product of the core of Schleiden's methodological ideas. Four years later, in the remarkable 158-page "methodological introduction" to his *Principles of Scientific Botany,* amounting to half of the first volume of the work, he developed his ideas under the heading: "Methodological Foundation."[24] (To place it within its period, remember that Mill's *Logic* appeared in the following year!) Sachs tells us that "the difference between this and all previous textbooks is the difference between day and night."[25] At one fell swoop we are removed from the preoccupation of previous botanical texts with systematics and purely empirical or observational matters. In its pages we find a systematic presentation of morphology, anatomy, and physiology, but the center of gravity is based on the chemistry of plants and on the form and growth of the plant cell. For us here, however, what is of importance is the Methodological Introduction.

This Introduction has had a peculiar fate. When the *Principles* were translated into English in 1849 (from the second edition), the Introduction was omitted altogether, and replaced by a superficial summary of some of its contents in two pages. In Germany, in contrast, methodological sophistication seems to have had a stronger hold. When subsequent editions of the *Principles* failed to keep up to date, students, though advised to refer to more recent texts for the scientific subject matter, were enjoined to continue the study of Schleiden's Introduction![26] (It is also of interest to note that when the English version of the *Principles* was reprinted as recently as 1969, it appeared again without the Introduction!) This Introduction is a veritable storehouse of stimulating material on methodological questions, with, among other things, an extremely sophisticated approach toward induction. To be sure, Schleiden's induction is not at all that of Bacon and Mill, or that of later neo-inductivist versions. It is rather more like the induction of two other nineteenth-century philosophers of science, E. F. Apelt's in Germany, a follower and sometime associate of Schleiden, and to a lesser extent, Whewell's in England. The difference is marked by Apelt when he speaks of "rational induction," which employs the Kantian conception of regulative principles or "leading maxims," as distinct from the induction of Bacon and Mill, which Apelt calls "empirical induction."[27]

The difference does not always seem to be appreciated. Thus J. Lorch, in his introduction to the 1969 reprint, remarks that Schleiden's methodological introduction "has done much to perpetuate the view that induction is the only key as well as the guarantee of a true science."[28] The critical tone of voice suggests that the writer believes Schleiden to be employing the classical induction of Mill and his followers. On the contrary, Schleiden's views on induction, like his general philosophical position, are closely based on those of Jacob Friedrich Fries, who in the first three decades of the century developed the basic Kantian approach to science in a number of systematic treatises, which are meant to constitute a distinct and conscious alternative to the *Naturphilosophie* of the Schelling and Hegel tradition.[29]

In their main outlines, and as far as concerns his philosophy of science, Fries's philosophical doctrines follow the Kantian teaching, though he presents them in a more systematic, less tentative, and more dogmatic fashion, and takes little account of the difficulties and ambiguities of Kant's own procedure. In his approach toward the "physics and metaphysics of nature," Fries follows the Kantian division into general and special principles, which in turn are divided into "constitutive principles" and "heuristic maxims." The constitutive principles embrace a heterogeneous set of mechanical laws and mathematical and kinematical foundations, in each case treating of a general and a special instance; the special instance, for example, treats of the "metaphysics of external nature."[30]

The basic postulate in all this is that all qualitative experience must be converted into metaphysico–mathematical universal laws in order to yield a doctrine of scientific knowledge.[31] Evidently Fries is here following the general approach of Kant's *Metaphysical Foundations of Science*. On the other hand, more explicitly and systematically than Kant, he operates with a set of particular heuristic principles which formulate the condition that all external experience of things must be regarded as a series of changes, at the bottom of which there are physical processes due to the conflict between fundamental forces.[32] The whole of these processes, Fries holds, must be regarded as circular in kind, and he expresses this as the principle that nature is an "organised whole."[33] It is noteworthy that this definition applies both to the realm of inorganic and to that of biological phenomena. Fries does admit, however, that treating biological phenomena as a type of physical phenomena presents difficulties, since next to nothing is known about the origin of their peculiarly organic aspect. Seeing thus that a "rational physiology," i.e., one involving a metaphysico–mathematical treatment, is so far beyond the realm of possibility, he holds that we must fall back on general systematic description, involving a qualitative account in terms of phenomenal forces, primarily using the concept of "stimulus" (*Erregung*).[34]

In general, then, everything should be reduced to the relation between fundamental forces. Where this is not possible, we can only apply the general logical procedure of science, i.e., a systematic description, involving experimentation.[35] This general methodology is formulated by Fries in three "heuristic maxims of judgment": 1) the maxim of systematic unity, 2) the maxim of the extension of knowledge, 3) the subsumption of the particular under the general,[36] all closely resembling Kant's maxims enunciated in the first *Critique*.[37]

In the following year, in his influential *Wissen, Glaube und Ahndung* (1805) Fries summarizes this general approach once again. He emphasizes that while the realm of mental phenomena (of the "mind" [*Geist*]) is altogether excluded from interaction with matter, belonging as it does to an altogether different order of the appearance of things, the whole of the physiology of organisms falls under the type of theorizing whose primary aim is the expression of phenomena in terms of physico–mathematical laws. He rejects the received tradition of a separation of physics and chemistry from biology.[38] In consequence, and here we meet a definitive break with Kant, any special teleological approach toward biological phenomena is rejected.[39]

As in much else, in this point too, Schleiden follows suit. His antiteleological sentiments are perhaps most clearly expressed in a somewhat later work, his *Studies,* where he writes that

teleology or the doctrine of the purposefulness of nature in its correct application is now [1854] put aside by most of the important scientists. . . . Certainly, its correct application presupposes that the fact which it wishes to connect in accordance with teleological concepts, has already first been scientifically established separately and through its causal relation as ground and consequence.

And to the imaginary criticism that this seems to be in conflict with the teaching of Kant, he replies that one needs by no means to follow slavishly every relatively unimportant aspect in the teaching of a great man.[40]

Apart from Schleiden and Apelt,[41] I do not know how far these methodological ideas exerted any sustained influence on subsequent philosophers of science.[42] Apelt operates with the basic Friesian ideas (as he acknowledges explicitly in his *Theorie der Induction*),[43] with its distinction between the metaphysical principles of physics, yielding *a priori* laws, and the empirical application of these laws. The specifically inductive aspect of this process ("rational induction") is also caught in the notion of "leading maxims," clearly identical with the "heuristic maxims" of Fries.[44]

A few years ago, in an article on the biological sciences in the nineteenth century, Everett Mendelsohn asked "whether Schwann, the cautious mechanist, was influenced by Oken, the romantic philosopher of nature," and more generally, by the *Naturphilosophie* of Schelling.[45] The question arises because of Schwann's aim "to demonstrate the identity of the principle of development of plants and animals and to reduce the whole process to one governed by physical–chemical laws."[46] Now it is true that Schelling's *Naturphilosophie* does operate with the idea of a universe as a cosmic organism, thus incidentally introducing the idea of unification of inorganic and organic realms. If one studies carefully Schwann's language and general approach, however, he cannot avoid the conclusion that Schwann's sources lay more in the direction of the Kantian philosophy of Fries, together with an eclectic mishmash of Cartesianism, Leibnizianism, and atomism.[47] Moreover, Schleiden and Schwann were close friends during their Berlin period of the 1830s. Though positive proof from Schwann's philosophical jottings does not appear, it is possible that he may have come under the influence of his associate's general philosophical views.[48]

On a similar tack, we may also remark here that in Owsei Temkin's study of the forerunners of Darwin during the 1840s and 1850s, the discussion of the contributions of the school of "analytical mechanics" (including Schwann, Lotze, Helmholtz, and du Bois-Reymond), as well as that of the "metaphysical mechanists" (Vogt, Moleschott, and Büchner),[49] again only emphasizes their joint opposition to *Naturphilosophie*. Temkin pays little attention to the fact that the mechanistic approach was fed not only from the Cartesian or the atomistic tradition but also from Kant and Fries. Besides, in all this, Schleiden, together with Schwann and Lotze,[50] rather antedates the popular wave of the universal application of the concepts of law and force, and the idea of the reducibility of the organic to the inorganic approach of physico–chemistry. Nor should it be forgotten that their views are also just prior to the discovery of the "interaction of natural forces" (conservation of energy), which did so much to popularize the doctrines of this school.

III

To return to Schleiden, we have noted the importance, not to say pregnancy, of the principle of unity in his central paper, the "Contributions." Now this approach actually arises very naturally out of Schleiden's general account of induction, as interpreted by him. Much in the spirit of Fries, he enumerates two modes of entry to a science that includes both laws and matters of fact.

1. We may develop the forms of rational knowledge and from this "derive problematically the possible laws (as done in [Kantian] philosophy of nature), and then apply the laws thus found to the fact. . . ."[51] In this context, apart from principles of rational mechanics, Schleiden cites the example of the law of gravitation, apparently being under the impression that for Kant this is derivable *in toto* from concepts of the possibility of material nature. Evidently Kant was as capable of misinterpretation then as he is now.

This is Schleiden's version of Kant's and Fries's "metaphysical foundations of science." Its object is to eradicate the failure of "pure empiricism . . . to hide behind the ambiguity of its vague and deficient abstractions, whose classification should rather be expected from philosophy alone."[52] The whole thing is really a program of philosophical analysis, or explication, as Schleiden makes clear:

Words like Organism, Life, Nisus, Soul, etc., are simply blanket terms for ignorance or lack of clarity [he writes], and here only a sound philosophical education can tell us: "This is the correct path of abstraction; through it we are led to such and such specific differences, with which, as signs, we may then connect just that specific term."[53]

2. The second entry to science is from the side of the facts, which we begin by analyzing, after which we seek to discover the conditions of their existence, subsequently moving to still higher universal and simple conditions.[54]

For the most part, Schleiden concedes, the first of the above ways cannot be employed because of the complexity of the phenomena, the exception being elementary kinematics and mechanics. Hence we are almost always dependent on "the inductive method," albeit in the sense of Apelt's later "rational induction." In all such cases, the "a priori developed laws of a philosophy of nature" are nothing

but "leading maxims, rules, in accordance with which we judge pro-
gressively the admissibility of hypotheses. . . ."[55]

When we turn to the organic realm of botany and zoology, these
considerations apply with particular force. Although our aim here,
too, must be to reach the standpoint of mathematical physics, with
its quantitative laws of force, we are as yet far removed from this
ideal, especially as regards the specific concept by means of which
we find that this subject has to be articulated, viz. the concept of the
formative propensity (*Bildungstrieb*).[56] Lest it be thought that this
is just a slide back into earlier conceptions of an intrinsic "life-force,"
Schleiden spends considerable effort on making it clear that this concept must not be understood as a purely tautological answer to the
problem of life, on the lines of the traditional notion of such a "life-
force,"[57] which wrongly postulates an essential difference in kind
between inanimate and animate,[58] but as the aim to express the phenomena of biology through laws of force which define their "mode
of action" and "lawlikeness."[59] "In the terrestrial organisms there
rules absolutely no different (still less higher-order) lawlikeness,
than in the purely mathematically and mechanically constructed
solar system."[60]

Schleiden is of course quite aware of the difficulties surrounding
this notion of a formative propensity.[61] That is, "formative propensities," like all "natural propensities" (*Naturtriebe*), of which they
are a special case (but unlike "fundamental forces"), suffer from the
disadvantage that here the "universal aim of science," consisting in
"the mathematical construction of the forms of interaction, and
similarly hence of the self-maintaining and formative propensity of
plants," is not yet in sight. Moreover—and this is important—we here
find the reason why, according to Schleiden, this notion of a formative propensity must be regarded, not as an empirical fact, but only
as a "leading maxim" which "thereby formulates a definite task" for
this branch of science.[62]

This idea of "leading maxims" we have already met as an important aspect of the formulation of the inductive method as understood by the school of Fries. The concept is important for Schleiden;
it rebuts the accusation that his formative propensity—as Sachs (who

failed to grasp Schleiden's logic) was later to object[63]—is only the life-force in disguise. The difference is that a leading maxim evidently possesses (so far at least) no existential import. It is interesting to find that the same remark applies to another of Schleiden's special leading maxims, that of "developmental history," which, quâ leading maxim, involves for him no claim to truth, nor, for that matter, to falsity. Sachs's response is again interesting. Whereas he complains that the reduction of the, for him, otiose concept of formative propensities to the status of a "leading maxim" was "irrelevant," now, per contra, with his whiggish consciousness of the contrasting "success" of the theory of descent, he complains that it is "superfluous to present" such a notion "as a 'maxim' in Kant's use of the word, instead of showing that the history of development enters naturally of itself into the investigation."[64] Perhaps Schleiden is here ahead of his time, for only in our own day has it been generally acknowledged that the status of developmental approaches, or—to cite another example—of uniformitarianism in geology, is best presented as that of leading maxims.

In defining the logical status of this notion, Schleiden follows the guidance provided by Kant and Fries, although he refers here to the latter alone explicitly. The account is to be found in the final section of his Introduction (Section 4: "Of induction in particular").[65] We move, he tells us, from facts to theory mainly through induction, hypothesis, and analogy.[66] Now all these yield only probable conclusions, though—as Schleiden explains, with a reference to Fries—it must be understood that the notion of probability here involved is "philosophical" and not "mathematical." In consequence, the fact that we do not suspend judgment, but on the contrary assign to it a certain degree of rational belief, Schleiden tells us, "lies in the nature of cognitive reason which everywhere demands unity and coherence in its investigations."[67] We have here at long last an explanation of that brief and mysterious reference to unity at the start of the 1838 paper with which we began.

It follows, Schleiden continues, that inductive conclusions are

only valid in so far as they agree with the whole of our faculties that lead to scientific knowledge [*Erkenntniskraft*], and with those principles which are derived from the latter. In fact, everywhere we presuppose unity and lawlikeness as given, and for preference choose what agrees with this presupposition.[68]

Schleiden uses here the Kantian notion of "reflection," which

employs the universal principles of reason, not as rules *under which* we subsume anything, but *in accordance with which* reflection determines its judgment, and this judgment is therefore only valid if it is arrived at in full agreement with the totality of rational knowledge.[69]

Certainly, "hypothesis, induction and analogy," without the guidance of maxims, must "lead to error and nonsense."[70]

IV

So far, this teaching is very much in the spirit of Kant, though Schleiden ties the use of "leading maxims" more intimately to the processes of induction, analogy, and hypothesis. The next step is more revolutionary, for at this point, going beyond Kant and proceeding more systematically than Fries, Schleiden divides his list of leading maxims into two groups, general and special. The former include the by now traditional tetrad, the maxims of unity, multiplicity, objective validity, and economy.[71] Only the third of these requires a brief explanation here. It simply says that just as the universal cannot arise out of the particular, so the explanatory principle cannot arise from the particular explanandum; rather, the particular is subject to universal conditions or determinations. This somewhat puzzling language probably signifies simply the Kantian contention that hypotheses and theories can be granted only as the spontaneous offspring of "subjective" theorification; the theoretical realm is not "given," but "set as a task"; it is no more than a "projection."[72] Moreover, with a long quotation from Fries's *System of Logic,* we get again the well-known Kantian criteria that a satisfactory explanation permits of no *ad hoc* hypotheses, that it allows no exception to its predictions, and finally that the agent of the explanation must have been shown to be "possible."[73]

At this point, Schleiden's methodology takes an interesting turn. Just as general transcendental logic and special mathematical philosophy of nature suggest the leading maxims of the general kind (*an die Hand giebt*), so we "must derive" (*abzuleiten haben*) the special leading maxims from the individual disciplines of natural science.[74] This is puzzling. Surely Schleiden cannot mean that these maxims can be straightforwardly derived either deductively or inductively? The answer to this must certainly be negative. On the other hand, it is not made very clear what logically clinching moves do determine the choice of the relevant maxims—maxims which are meant to provide a paradigmatic point of view and act as principles of acceptance and rejection of possible hypotheses.

We are reminded of what is implied in Sachs's objection: with what right do you choose your leading maxims, if they are not simply inductive or hypothetico–deductively determined principles of explanation? It seems to me that Schleiden does not face up to this question too clearly, perhaps because he is led by the conviction that the leading maxims of the general kind have a truth-value-less logic into the belief that a similar position is plausible (and incidentally, as I have already said, safe as well) when we start operating with leading maxims of the special kind. Hence no particular logical safeguards would seem to be called for here either.

Let us see what special considerations Schleiden actually does put forward in favor of these maxims. He begins with the reminder that the hypotheses of every special science must satisfy, as far as possible, the criterion of consilience, or better, of coherence with extra-theoretical contexts. They must not be inconsistent with other branches of science. In practice, for botany this means that its inquiries should avail themselves of the standpoints of physics and chemistry. In this way, such extra-theoretical disciplines acquire a special significance as auxiliary sciences, namely, by supplying, with the supplementary aid of induction and hypothesis, the means to bring to light (*nennen*) the leading maxims for the advancement of botanical knowledge.[75]

The leading maxims (Schleiden does not tell us what they are) that thus arise are insufficient, however. We need to descend further in order to specify the highest maxim, unity (the first of the universal maxims, it will be remembered) in a more determinate way. The maxims which then result, he insists again, are principles of unity in virtue of which we do not construct the science, but through which we let ourselves be guided in our inductions. And at this point comes the crucial pronouncement: In order to ascertain the special leading maxims, they "must of course be derived [*ableiten*] from the nature of the object of the discipline concerned, i.e. botany; and this we do by collecting the general and altogether assured facts [of the science] and expressing them as a rule."[76]

This extremely subtle position clearly bears a distinct resemblance to many accounts that have been evolved in the more advanced, and historically informed, methodologies of recent vintage,[77] although, as I showed in some detail in section I above, Schleiden does not give sufficient recognition to the way in which such leading principles may affect the "descriptive" subject matter of his science. Certain well established facts—and it will be appreciated that by "facts" Schleiden has in mind both observational and theoretically informed results, processed with a physico–chemical terminology as its basis—are adopted as rules for the further organization, selection, or rejection of the phenomena, past, present and future, of the science concerned. Important here is the insight that the most general theoretical standpoints are not so much "facts" themselves, but things that have been converted into principles bearing the status of rules.

V

As far as botanical maxims are concerned, Schleiden proposes two: 1) the maxim of developmental history, and, 2) the maxim of the independence of the plant cell. The latter notion includes in some way, as a special case, the idea of formative propensity, discussed previously.

The rule corresponding to the first maxim is that every hypothesis, every induction in the science of botany must be unconditionally rejected, unless its orientation is guided by the history of development.[78]

The approach which Schleiden here advocates, that the nature and significance of an organism can only be understood through the history of its development, he illustrates by the history of systematics and the evolution toward natural systems of taxonomy. We cannot grasp the present state of a plant (and hence cannot classify it) without studying its evolution from previous states. "Development" for Schleiden means the aspect of growth of a plant. The "whole series of intermediate steps must have become an object of observation," he tells us, for only then can we have "gained a secure foundation for induction, so as to derive the lawlike regularities of its changing states"[79]—a good example of his conception of the relation between regulative maxim and induction.

The rule expressing the second maxim makes clearer what is intended:

every hypothesis, every induction, must be rejected unconditionally, if it does not have as its aim to explain the processes taking place in the plant as the result of the changes which take place in its individual cells.[80]

The analytic principle thus enunciated emphasizes physico–chemical studies of the individual cell.[81] Cell physiology must precede the physiology of the whole plant.

If Schleiden was an early member of the school of "analytical mechanics," though approaching it through Kantian channels, we now see him also, together with Cotta and Robert Chambers, as a distinguished forerunner of evolutionary theory.[82] Indeed, according to Rádl, through the new method which Schleiden introduced into biology, "he prepared the minds for the acceptance of Darwinian philosophy,"[83] although no doubt there were many minor figures as well.[84] It is worth remembering that even the celebrated and influential *Vestiges of Creation* (1844) was not published until two years after the first appearance of Schleiden's *Principles*.

The developmental and cell maxims are employed in a number of Schleiden's writings. Already in his "Contributions" Schleiden had centered the discussion on the concept of growth and development:

The imagination obtains ample latitude for the explanation in every case of the generation of infusorial vegetable structure, even without the aid of a *deus ex machina* (the *generatio spontanea*).[85]

In thus focusing attention on embryological growth, Schleiden again is far ahead of his time, anticipated only by a few figures such as Meckel and von Baer, who proceed of course without attention to cell theory.

A few years later, in his *Popular Lectures on the Plant* of 1848, we find him applying the idea of geological sequence also to the realm of fauna and flora. The mode of argument should again be closely scrutinized. Schleiden hypothesizes that at any early period, in virtue of forces still extant, but under conditions no longer existing, there probably arose "the first germs [*Keime*] of organic beings."[86] A little later he further canvasses the idea that the different species may well have developed from a single cell and its descendants through variation and gradual formation of kinds, which eventually formed the different species.[87]

We are not interested here in the purely historical question of Schleiden as a forerunner of Darwin. Rather we note that our "leading maxim" is based, on the one hand, on the kind of surmises that Newton would have called "Queries," as well as on certain extensions of observational considerations. On the other hand, the result is not allowed to stand as a putative hypothesis, let alone established fact, but is converted into a principle of research.

Schleiden's method of approach to evolutionary theory has been characterized by Tschulock as follows:

As regards its logical nature, the thesis of organic continuity is an example of those categories of propositions which should be counted among the "incomplete inductions." We do not know in *all* cases by experience that given organisms are produced by their parents. But in the process of generalising the numerous cases with which we have become acquainted with certainty to a proposition in universal form, we base ourselves in turn on a postulate, on a demand of reason. We say: one and the same result can hardly have been produced in two so different ways. Only by appeal to such a postulate does there result an empirical principle of widest application: "every organised being originates by way of reproduction."[88]

This is another way of answering the question posed earlier concerning the relationship between induction and leading principle. But it omits the "regulative" force of the resulting "postulate" (as Tschulock calls it), which for Schleiden was an important aspect of his "leading maxims."

Schleiden is one of the earliest nineteenth-century figures to usher in the period of the "mechanical approach" to nature, though not as "fact" but as "leading principle." Furthermore, he provides an alternative both to the purely observational and to the static classificatory approaches of biology. Finally, he adds another voice in the rising opposition to the doctrines of the *Naturphilosophen,* especially Schelling and Hegel, whom Schleiden, in common with non-Kantian mechanists of his time, never ceases both to attack and to cite for the errors of their approach. What is remarkable is Schleiden's appreciation of the central position of method in the sciences, and most especially in botany. Discussing the primitive state of this science in his own time, in comparison to the heroic achievements of "Galileo, Kepler and Bacon,"[89] and quoting the "sad comment" in Whewell's *History of the Inductive Sciences*[90] on this state of affairs, he ends by proclaiming that there is an absolute "justification and necessity, in the case of every scientific activity, to enquire first of all into its method, to test it, and according to the result of this test alone to praise or to dismiss its labours."[91] And "faithful to the spirit of my great teacher Fries," we must "demand that every scientific discipline should advance exclusively in accordance with the inductive method"[92]–though to be sure, with "induction" interpreted as "rational," in the sense here indicated.

Schleiden does not mince words when he turns to the attack on Schelling and Hegel. With characteristic overstatement, he proclaims that Fries is the only one to have taken the trouble to unravel the confusions that have infested philosophy since Kant, and the only one even to grasp the true task of philosophy and its solution by Kant. In the face of Fries's strictures, Schleiden asserts indignantly that Schelling had simply remained silent, while Hegel had responded by insult with his "impertinent remark" that "Fries is the commander-in-chief of those who do not think" (*"dass Fries der Heerführer aller Nichtdenker sey"*).[93]

In conclusion, we may thus appreciate the close relation that existed here between philosophical history, logical principles, methodological precepts, and the living history and inductive results of science, in a man who was no "mere scientist," still less a "mere philosopher of science," but an honored figure in the growth of science. And we are most impressed by his enunciation of a complex methodology far in advance of his time, matched only by an equal anticipatory insight into scientific developments that were to become

ruling dogmas not more than one or two generations later. To summarize, Schleiden clearly distinguished between the purely inductive
strand of science, on the one hand, and the need consciously to
formulate a conceptual frame—the physico–chemical terminology of
force—in a branch of science that had not hitherto employed such a
frame, extending to botanical inquiry Kant's celebrated slogan that
"in every branch of natural science there is only as much science
proper as there is mathematics therein."[94] He enlarged the Kantian
conception of regulative principle in a most interesting way, through
his development of the notion of specialized leading maxims, which,
while based on inductive suggestions, may leap ahead, and through
their direction-giving power yield structure and content for the
growth of a science. Schleiden thus provides a very good illustration
of the contention advanced by Merz that it needed more than "the
exclusive development of the methods of exact research" to produce
the reform of biology in nineteenth-century German science, but
that this required in addition "the philosophical, historical, and critical spirit which formed the peculiar characteristic of German
thought" at that time.[95]

VI

Nothing like an exhaustive analysis of Schleiden's Methodological
Introduction could be provided in this paper. In order to help the
reader gain some idea of the complete system, I will append an
analytical table in which Schleiden's somewhat sprawling and discontinuous presentation is displayed in systematic fashion. The following then are Schleiden's chief principles or "norms" of scientific
investigation, with special reference to the methods of botanical
investigation. The whole presentation provides an interesting
illustration of the triadic methodological classification that I have
tried to sketch in some of my own writings, referred to in notes 37
and 77.

I. The mechanistic analysis of nature.
  a. Explication of concepts.
    i. All bodily substance is matter, defined as what fills space
       in virtue of a special motive force. Matter is not atomistic
       but "dynamic," the source of forces subject to exception-

less laws in mathematical form. Some of these laws are basic ("constitutive") and derivable from conceptual considerations, subsequently to be applied to the experiential domain. The "metaphysics of nature."

ii. The distinction between fundamental forces (*Grundkräfte*) and natural propensities (*Naturtriebe*). The latter are not a *special* substance or force, of a different kind, e.g., "life-force," assumed hypothetically, but must always be derivable (if only in principle) from a mathematical law of fundamental forces (constructed by the metaphysics of nature) and the geometrical relationships of movable masses in space. Natural propensities concern the contingent or particular aspect (e.g. boundary conditions) of all natural processes, e.g., the propensity for planetary orbits to have a particular inclination to the ecliptic.

iii. Natural processes are: gravitation, processes involving wave motion, electromagnetic phenomena, chemical phenomena, and morphological processes. The latter involve a special case of natural propensity, viz. formative propensities (*Bildungstriebe*). These also are combinations of fundamental forces, though again still unknown.

b. The principle of analysis of wholes into parts.

c. All determination is by form and law.

d. The ideal of explanation is mechanical, the reduction to physical and chemical processes, and of all change to motions according to mathematical laws, in terms of interaction between masses at a distance or through touch.

II. The inductive method. ("Empirical induction.")

a. Instead of axioms and the synthetic method, of hypotheses and principles, e.g., life-force, we must start with immediate sensory certainties; advance from singular facts which are then ordered by subjecting them systematically to a hierarchical system of laws, up to the highest concepts and laws.

b. In the sense of a., *every* proposition is an incomplete theoretical whole.

c. Individual facts and objects are to be analyzed to determine their original parts and conditions, thence to higher-level and simpler conditions, up to the highest. These conditions must belong to the physico–chemical realm.

d. The scientific procedure moves in stages.

    i. Systematic description of nature (morphology, taxonomy).

    ii. Teleological description of nature (structure and function).

    iii. Combining observation or inspection. Comparison of related experiences to discover the law under which they stand.

    iv. Theoretical experimentation (and observational methods). Examination of underlying experimental methods, e.g., optics microscopy.

    v. Mathematical theory (hylology). Physicalist treatment of nature, involving basic constitutive principles in mathematical form.

III. The employment of regulative principles ("leading maxims"), assuring of systematic unity and accordance of inductions. ("Rational induction.")

a. General maxims.

    i. Maxim of unity: all knowledge is "reducible" to principles.

    ii. Maxim of diversity: laws and rules require the greatest possible range of applications to individual facts.

    iii. Maxim of objective validity: the particular is subject to universal conditions.

    iv. Maxim of economy: principles are not to be multiplied without necessity.

b. Special maxims.

    i. Maxim of developmental history: the rule, so to observe a continuous sequence of states, from fluid state to cell, and from the latter to the composite plant, that there is no concealed gap which would need filling by means of hypotheses.

    ii. Maxim of the independence of the plant cell: the life of the plant is to be explained as the result of changes taking place in its individual cells.

NOTES

1. M. J. Schleiden, "Contributions to Phytogenesis," in Müller's *Archiv für Anatomie und Physiologie,* Part II (Berlin, 1838), pp. 137–76; in Taylor's *Scientific Memoirs,* vol. 2, Part VI, trans. Francis (London, 1841), pp. 281–312; republished together with Schwann's *Researches* (see note 3) by The Sydenham Society (London, 1847), pp. 231–63.

2. "Contributions," p. 251.

3. Theodor Schwann, *Microscopical Researches into the Accordance in the Structure and Growth of Animals and Plants* (Berlin, 1839); trans. Henry Smith (London: Sydenham Society, 1847).

4. *Researches,* p. 192; cf. pp. 190–191; also Schleiden, "Contributions," p. 249.

5. *Principles* (see note 9), p. 214. Cf. pp. 74–5: "It is first of all important to establish inductively . . . the task, to reduce organic processes to purely physical reactions. . . . Through such a presentation we obtain a well-nigh irrefutable induction for the truth that in organisms there exist no other fundamental forces whatsoever but those [that are found] in inorganic nature."

6. *Researches,* p. 187.

7. For some of these connections, see for instance Owsei Temkin, "The Idea of Descent in Post-Romantic German Biology: 1848–1858," in *Forerunners of Darwin: 1745–1859,* ed. Bentley Glass et al. (Baltimore: The Johns Hopkins Press, 1959), pp. 323–55.

8. Cf. *Principles* (see note 9), 2nd ed., p. 197. In the third edition, p. 204, there is no mention of the nucleus at this point.

9. M. J. Schleiden, *The Principles of Scientific Botany, together with a Methodological Introduction as a Primer for the Study of the Plant,* 2 vols. (Leipzig, 1842–43); 2nd ed., 1845–46 (subtitled: *Botany, treated as an Inductive Science*); 3rd ed., 1849–50; 4th ed., 1861; 2nd ed., trans. E. Lankester (London, 1849), without the Methodological Introduction; Facsimile of the London 1849 translation, with a New Introduction by Jacob Lorch (New York: Johnson Reprint, 1969). Unless otherwise noted, all references are to vol. 1 of the third German edition of 1849; all translations are mine.

10. *Principles,* p. 214.

11. *Researches,* p. 201.

12. M. Möbius, *Matthias Jacob Schleiden zu seinem 100. Geburtstage* (Leipzig, 1904), pp. 16, 29. Mechanistic principles were of course "in the air" of the early nineteenth century. Thus, J. R. Mayer, to whom we owe the earliest formulation of the principle of the conservation of

energy (1842) knew Schwann's "Microscopical Investigations," as he indicates in a reply to an inquiry of 1842 from his friend, the medical practitioner W. Griesinger. Griesinger's original letter is, moreover, of great interest here since it shows very clearly that by 1842 the anti-vitalist physico-chemical approach was common ground for many of the scientists of the time. Thus Griesinger actually refers to Schwann in connection with his own view according to which it would be a great "progress, wherever possible to deprive the processes taking place in organisms, of the mysterious mysticism of the vitalists etc., and to find for them something analogous or identical in the remaining material substance, to which we might suppose its organised aspects to be subject also. . . . The development and execution of a purely physical view of vital processes I consider to be the task of the physiology of our time. You will be acquainted with the splendid contributions to physiology which have been contributed by Schwann for instance." (*Kleinere Schriften und Briefe von Robert Mayer,* ed. J. J. Weyrauch, [Stuttgart, 1893], p. 197.)

13. Arthur Hughes, *A History of Cytology* (London & New York: Abelard Schumann, 1959), p. 39.

14. "Contributions," p. 247.

15. Julius von Sachs, *History of Botany (1530–1860)* (Munich, 1875), trans. H. E. F. Garnsey, rev. I. B. Balfour (Oxford, 1906), p. 324. All references are to the 1906 edition.

16. Ibid., p. 296.

17. *Principles,* p. 34. It is of course an open question whether the generalized notion of theory-ladenness does follow from the Kantian epistemology. I have discussed this problem in my "Inductivist versus deductivist approaches in the Philosophy of Science as illustrated by some controversies between Whewell and Mill," *Monist* 55:3 (July 1971): 343–367. Issue devoted to British philosophy in the nineteenth century.

18. Ernst Cassirer, *The Problem of Knowledge: Philosophy, Science and History since Hegel,* trans. W. H. Woglom and C. W. Hendel (New Haven: Yale University Press, 1950), p. 155.

19. Ibid. I have quoted Sachs, *History of Botany,* p. 188.

20. Ibid., p. 327.

21. Ibid., p. 323.

22. "Contributions," p. 231.

23. Ibid., p. 232. This is followed up by a lengthy disquisition on the meaning of the verb "to grow."

24. *Principles,* pp. 1–162.

25. Sachs, *History of Botany,* p. 189. There are many other such eulogistic praises to be found among Schleiden's contemporaries and

later historians. Cf. also Lorch's New Introduction to the *Principles* (see note 9), pp. xv–xvi, for the judgments of Unger and Goebel.

26. Cf. Möbius, *Matthias Jacob Schleiden*, p. 22.

27. E. F. Apelt, *Die Theorie der Induction* (Leipzig, 1854), p. 41. Cf. also Cassirer, *The Problem of Knowledge*, p. 147, for a reference to Apelt, in connection with Schleiden.

28. Lorch, New Introduction to *Principles* (see note 9), p. xviii.

29. Thus in *Schellings und Hegels Verhältniss*, which is something of a supplement to the Methodological Introduction provoked by a hostile review of his *Principles*, Schleiden writes:

In my methodological introduction I have explicitly presupposed for my considerations the point of view of the philosophy of Fries. . . . A just critique hence can only ask: Is the Friesian philosophy here correctly understood and correctly applied? . . . By the way I will only mention that I discussed the whole methodological introduction, sentence by sentence, with our Fries, so that I am sure that at least in essentials I have followed his sense.

(Quoted in Möbius, *Matthias Jacob Schleiden*, p. 24.)

30. Fries's principles are outlined in a number of writings including his *New Critique of Reason* (Heidelberg, 1807), *System of Logic* (Heidelberg, 1811), *System of Metaphysics* (Heidelberg, 1824), *Mathematical Philosophy of Nature* (Heidelberg, 1822). I have preferred to follow here one of his earliest works, the *System of Philosophy* (Heidelberg, 1804), which was actually completed by 1801, in order to show how far these ideas precede and are independent of the current doctrines of the *Naturphilosophie* of the period, and how far they also antedate the mechanistic approaches of the 1840s to 1860s. References to the *System of Philosophy* are taken from: *System der Philosophie als evidente Wissenschaft*, in Jacob Friedrich Fries, *Sämtliche Schriften*, ed. Gert König and Lutz Geldsetzer, Abt. 1, Bd. 3 (Aalen: Scientia, 1968). The pagination is that of the 1804 edition. The account of the "Metaphysics of External Nature" occurs on pp. 286–315.

31. Fries, *System*, pp. 286–7.

32. Ibid., p. 295.

33. Ibid., p. 295.

34. Ibid., pp. 313–15.

35. Ibid., pp. 283–4.

36. Ibid., pp. 258–9.

37. For Kant's theory of regulative principles, cf. *Critique of Pure Reason*, trans. N. Kemp Smith (London, 1953), pp. 542–9 (A657/B685–A668/B696). I have discussed Kant's regulative approach also in my *Metaphysics and the Philosophy of Science. The Classical Origins: Descartes to Kant* (Oxford: Blackwell, 1969; Cambridge: M.I.T. Press, 1970), pp. 506–16.

38. Fries, *Wissen, Glaube und Ahndung* (Jena, 1805); *Schriften,* pp. 92–3.

39. Fries, *System,* pp. 378–9.

40. M. J. Schleiden, *Studien* (Leipzig, 1855), pp. 156, 159.

41. Lorch (see note 9), p. xiii, quotes from Schleiden's judgment on Apelt, which shows that he had for him the highest regard. Besides, Schleiden and Apelt together with others, collaborated in 1847 to edit the *Abhandlungen der Friesischen Schule.* See also J. T. Merz, *A History of European Thought in the Nineteenth Century,* (London, 1896), 1:208 and n. 2, which cites an impressive list of names said to have derived from "the school of Fries."

42. But see the anecdote quoted in Lorch's Introduction (see note 9), p. xiii, from Schleiden's *Über den Materialismus,* concerning the mathematician Gauss's admiration for Fries's *Mathematical Philosophy of Nature.*

43. Apelt, *Die Theorie,* p. 171.

44. Ibid., pp. 53, 92–3. Apelt however added and made central another methodological notion, viz. abstraction, supposed to relate "metaphysical" laws of physics to the empirical phenomena. Apelt, particularly in this use of "abstraction," seems to have exerted some influence on the philosophy of Ernst Mach. (I have in my possession Mach's copy of Apelt's *Theorie der Induction,* copiously marked. Also Mach's frequent references to it in *Erkenntnis und Irrtum* testify to its interest for the younger scholar.)

45. Everett Mendelsohn, "The Biological Sciences in the Nineteenth Century: Some Problems and Sources," in *History of Science,* ed. A. C. Crombie and M. A. Hoskin, (Cambridge: Heffer, 1964), 3:42.

46. Ibid. See also Mendelsohn, "Schwann's mistake," which pays special attention to Schwann's physico–chemical approach to cell theory. (In *Proceedings of the Tenth International Congress of the History of Science* [Ithaca, 1962], 2:967–70.)

47. For Schwann's philosophical views see R. Watermann, *Theodor Schwann. Leben und Werk* (Düsseldorf: L. Schwann, 1960), pp. 132–4, 204–22. Any direct evidence for the claim of a Friesian influence is of course absent, apart from Schwann's reiterated belief that the basic aspect of matter is force (e.g., ibid., p. 207), and his rejection of the mechanical atom, in favor of the Kantian dynamical definition:

Every atom is a definite, altogether individual force which cannot be amalgamated with others and which acts in a given space, viz. the volume of the atom, with the exclusion of every other force. In this space however it acts in all directions, and through this there arises corporeality. (Ibid., p. 132)

Schwann's biographer finds this definition "surprising." It was however a commonplace for the Kantian school.

48. For the meeting at which Schleiden communicated to Schwann the gist of his theory of the plant cell, see Watermann, *Theodor Schwann* (note 47), pp. 98–100, 200.

49. Temkin, "The Idea of Descent" (note 7), p. 325.

50. For Lotze, see J. E. Erdmann, *Grundriss der Geschichte der Philosophie* (Berlin, 1866), 2:781–95. Note especially Lotze's *Allgemeine Pathologie und Therapie als mechanische Naturwissenschaften* (Leipzig, 1842), and his entry, "Life," in Wagner's *Handwörterbuch der Physiologie*, of the same period. In the former, Lotze seeks to show that what takes place in animate bodies differs from inanimate physical processes, not by virtue of a difference in principle in the nature and action of the forces that occur, but by the arrangement of their points of attack. In particular, he argues that the "life-force" is not so much a force, but rather the magnitude of the effect which is produced by the joint action of a number of separate physical forces under certain conditions. (Cf. Erdmann, *Geschichte,* p. 785.)

51. *Principles,* p. 40.

52. Ibid., p. 10.

53. Ibid., p. 10.

54. Ibid., pp. 40–41.

55. Ibid., pp. 41–42.

56. Ibid., p. 51. I use "propensity" rather than "impulse" or "nisus," in order to remind the reader of Schleiden's intention that the term *"Bildungstrieb"* should be used with "phenomenalist" intent. Still, it must be admitted that the terminology to a certain extent allowed Schleiden to have it both ways, since *"Trieb"* still smacks of the older types of special life-forces, plastic tendencies, etc. (For this reason, I have desisted from using the more plausible term "tendency.")

57. Ibid., p. 56.

58. Ibid., p. 57.

59. Ibid., p. 57.

60. Ibid., p. 61.

61. Ibid., p. 62; cf. pp. 49, 56. Cf. also Schleiden's reference back to Fries, Ibid., p. 74. "Formative propensities" are a special case of "natural propensities" (*Naturtriebe*). (See also my Analytical Table, I.a.ii–iii, for more details.) The relevant passages from the *Principles* are worth quoting here in full:

The forms of interaction in the physical world are determined by the fundamental forces [Grundkräfte] of matter in different substances and through the forms of aggregation. Here, only the following needs to be stressed: Gravitation, for instance, rules the continued existence of the whole machinery of the heavens with eternal lawfulness. However, this explains after all only its existence on the assumption of a purely geometrical relationship to space, through which is given only the continuance of motions in conic sections but not of tangential motion, and hence not the possibility of the origin of the orbits. This draws our attention to the unavoidable incompleteness of our knowledge of nature, which arises out of its finite limitation within the infinity of time and space. For side by side with all lawlikeness in the sequence of appearances there always remains the contingency of the mathematical composition which is altogether independent of the action of the fundamental forces. Now the relationships that arise from this, e.g., the number of planets in our solar system, the sequence of planetary distances, the inclination of their orbits, are the properly particularising element in natural processes, and hence, in order to distinguish them from the fundamental forces, we call all these forms of interaction, in their connection with the fundamental forces, natural propensities [Naturtriebe], so as to distinguish them from the former. The mathematical construction of the main kinds of natural propensities would hence here be the proper task of Natural Philosophy [Naturphilosophie]. For we should never directly assume, as explanatory foundation for a natural process, a special substance or a special force, but only a natural propensity, which allows of derivation [ableiten] from a mathematical law of the fundamental forces and a geometrical relation to space of the movable masses. (Ibid., pp. 46, 47)

The universal aim of science, to establish the validity of the hylological [materialist] world picture, includes the mathematical construction of the forms of interaction and similarly therefore both the propensity to self-preservation and the formative propensity [Bildungstrieb] in plants. Until now we are still so infinitely removed from a solution of this task that we can only posit it as a demand on science, and employ it in its cultivation as a leading maxim. The attempt towards a solution of this task depends on the prior solution of three other tasks. For first of all we must have satisfied completely the task specified in Section 6 [the establishment of a natural classification of plants in accordance with morphological, anatomical and physiological characteristics]. Secondly, we must develop the construction of formative propensities from the already completely solved processes of gravitation, right up to the independent action of the forces in the organism (which however in turn presupposes first a completion of chemistry and physics). Thirdly, the construction of the morphotic processes must have been successfully effected in the simplest case of the crystals, so that everywhere the formative propensities can be subsumed under the hylological world picture. (Ibid., pp. 68–9)

62. *Principles,* p. 69.
63. Sachs, *History,* p. 189.
64. Ibid., p. 190.
65. *Principles,* pp. 137ff.
66. For an illustration of this process, it is best to refer again to Schleiden's own creative scientific work, which—apart from the more philosophical influences—ought to be expected to yield relevant evidence for his methodology of induction. Let me, therefore quote some of the central passages of the "Contributions":

As the *constant presence* of this areola [nucleus] in the cells of very young embryos and in the newly-formed albumen could not fail to *strike me* in my extensive investigations into the development of the embryo, it was *very natural* that the consideration of the *various modes of its occurrence* should *lead to the thought,* that this nucleus of the cell must hold some close relation to the development of the cell itself. . . .

If we consider that there are undoubtedly many plants, among which the *Fungi* and infusorial *Algae* should probably be classed more especially, in which we are, as yet at least, totally unacquainted with the cytoblasts [nuclei], in consequence of their absolute *minuteness and transparency;* if we further bear in mind that the nucleolus in the cell-germ, even in the larger cytoblasts, frequently appears *immeasurably small,* or even entirely escapes the eye with the highest magnifying power; and, lastly, if we deduce from what has been previously stated, that nevertheless this *granule,* which can no longer be rendered perceptible, *probably furnishes* in the suitable medium a *sufficient cause* for the formation of a cytoblast which serves as an introduction to the whole formative process of the cells; then, indeed, we are *forced to confess* that the *imagination* obtains ample latitude for the *explanation in every case* of the generation of infusorial vegetable structure, even without the aid of a *deus ex machina* (the *generatio spontanea*). (pp. 233, 248–9; my italics)

In these lines we see clearly at work the formation of a hypothesis of development, under the joint weight of constant conjunctions, variety of circumstances, elimination of negative instances and the provision of a putatively descriptive account of a developmental sequence for selected individual instances.

67. *Principles,* p. 137.
68. Ibid., pp. 137–8.
69. Ibid., p. 138.
70. Ibid.
71. Ibid.
72. Cf. Kant, *Critique,* p. 535 [A647/B675].
73. *Principles,* pp. 138–9. For this notion of "real possibility" in Kant, see my *Metaphysics and the Philosophy of Science* (note 37), ch. 8, *passim;* also my "The Conception of Lawlikeness in Kant's Philosophy of Science" (in *Proceedings of the Third International Kant Congress* ed. L. W. Beck [Rochester, N.Y., 1970]; and in *Synthese* 23 (1971): 24–46). Briefly, a scientific theoretical concept possesses "real possibility" if a) it is consistent with general categorical principles; b) it emerges from a mathematico–metaphysical construction, on the lines sketched in Kant's *Metaphysical Foundations of Science.*
74. *Principles,* p. 140.
75. Ibid., pp. 140–41.
76. Ibid., p. 141.
77. See for instance the contributions by Howard Stein, Kenneth Schaffner and myself in *Historical and Philosophical Perspectives of Science, Minnesota Studies in the Philosophy of Science,* vol. 5 (Minneapolis: University of Minnesota Press, 1970), and my "Gravity and In-

telligibility: Newton to Kant," in *The Methodological Heritage of Newton,* ed. R. E. Butts and J. W. Davis (Oxford: Blackwell and Toronto University Press, 1970), pp. 74–102.

78. *Principles,* p. 146.

79. Ibid., p. 144.

80. Ibid., p. 148.

81. Ibid.

82. Cf. Temkin, "Idea of Descent," pp. 332ff., 345.

83. Emil Rádl, *Geschichte der Biologischen Theorien,* Part II (Leipzig, 1909), p. 65.

84. Cf. Milton Millhauser, *Just Before Darwin* (Middletown, Conn.: Wesleyan University Press, 1959), p. 72.

85. "Contributions," p. 249. (See note 66 for the context of this passage.)

86. M. J. Schleiden, *Die Pflanze und ihr Leben* (Leipzig, 1848), p. 257.

87. *Die Pflanze,* p. 271; cf. pp. 268–9.

88. S. Tschulok, *Deszendenzlehre* (Jena, 1922), p. 136.

89. *Principles,* p. 6.

90. William Whewell, *History of the Inductive Sciences,* vol. 3, 3rd ed., (London, 1857), p. 217.

91. *Principles,* p. 17.

92. Ibid.

93. Ibid., p. 20.

94. Kant, *Metaphysical Foundations of Natural Science,* trans. E. B. Bax, (London, 1883), Preface, p. 140.

95. Merz, *European Thought,* 1:216.

3 Whewell's Logic of Induction

ROBERT E. BUTTS
University of Western Ontario

I. WHEWELL'S TWO THEORIES OF METHOD

Although the literature on nineteenth-century methodology includes some discussion of the Mill–Whewell controversy over the nature of induction, Whewell's own theory of induction has not received much attention. Especially lacking is any attempt at a philosophically interesting reconstruction of his logic of induction against the background of recent discussions of problems of inductive logic. Like Herschel, and unlike Mill, Whewell did not attempt a direct solution of the so-called "problem of induction," perhaps because he and Herschel did not think that such a problem existed. This omission from his work—if it be one—must have made his theory of induction somewhat less interesting philosophically than Mill's, and so Mill's theory achieved some stature in late nineteenth-century discussions of induction, and those of Herschel and Whewell became all but forgotten until quite recently.[1] This historical situation is unfortunate for two reasons. The ascendancy of Mill's way of dealing with problems of induction obscured the historical derivativeness of much that seemed novel in his *Logic.* For example, Mill's canons of induction are first set forth in Herschel's *Preliminary Discourse,* and much of Mill's understanding of actual science is derived from Whewell's historical treatment of inductive science. But more important, emphasis upon Mill as *the* Victorian philosopher of induction distorts the philosophical and historical picture of the development of nineteenth-century British scientific methodology. Herschel and

Whewell were scientists writing about science; their forebears were Aristotle and Newton. Mill was a philosopher using scientific examples to help in the solution of philosophical problems; Hume was his progenitor.

It is encouraging that scholars are at last beginning to attend to the fact that Victorian Britain was astonishingly rich in discussions of induction, so much so, in fact, that one might want to suggest that some of the seeds of our own contemporary understanding of scientific methods were germinating in the halls of Trinity and St. John's in Cambridge and elsewhere during the period 1840–1860. In this paper, I hope to add some evidence for this suggestion, and I hope also to be able to further my campaign of restoring Whewell's work on methodology by showing its philosophical novelty and its historical pre-eminence in Victorian thought about science. To attempt to accomplish both aims, I will examine in some detail Whewell's inductive logic, or rather his inductive logics, for I believe I can show that two different, though intermingled and confusedly related, theories of induction are involved in Whewell's work. Briefly, Whewell's two theories of induction are: 1) the completely non-novel hypothetico–deductivist account of the justification of scientific results, and, 2) the strikingly original (in the British methodological tradition) theory that the task of inductive logic is to generate a rule or rules on the basis of which theories are accepted or rejected, where such rules receive whatever justification they have from outside hypothetico–deductive systems of science. I am not suggesting that Whewell had nothing new to say about the classical hypothetico–deductive model. Indeed, it is the context of his attempt to explicate this model as a model for both discovery and justification that his second theory enters the scene, with the entrance largely resulting from the application of the commingled categories of Kantianism and classical British empiricism that formed Whewell's philosophy.

To understand the rudiments of Whewell's inductive logic just six of his concepts are required: 1) colligation via the superinduction of ideas on facts, 2) prediction, 3) consilience of inductions, 4) simplicity, 5) successive generalization, and 6) induction as demonstrative inference given the role of deduction in science. To understand his full view of induction many more ideas than these are needed—

chief among them the concept of "exemplification of a law" and that of the relativity of facts and theories.

II. SCIENCE AS INTERPRETATION

Whewell thought that the goal of science was to interpret nature, not to catalogue or describe it.[2] The goal of inductive logic is to show that science is the right interpretation. Science is not self-critical. It cannot use the methods of science to justify the use of those methods. We are at square one: science is interpretation. The goal of science is not to devise compendious schemes for summarizing facts, nor is it the linguistic mirroring of the extra-linguistic. Science is explanation; it is explication of what nature means. We can already get some good idea of how the strategy of the game will develop. For to interpret something is to try out some conceptual scheme; interpretation begins with the imposition of some concepts on that which is to be understood. The imposed scheme will either illuminate its object or it will not, it will either help us to see the object in a different way or it will not, it will either get us to attend to the familiar because we now recognize in it that which is unfamiliar but interesting, or it will not. There is, of course, no special magic in the visual imagery that Whewell uses over and over again; but though it may lack magic, such imagery is vital to Whewell's strategy. The aim of science is neither power nor control, but the attainment of a certain kind of conceptual seeing. It is no accident that the favored Whewellian metaphors are Platonic; science is just exactly that which will get us out of the cave.[3]

However, if at square one the strategy is clear, the ever-present tensions within that strategy are also clear. Of course the aim of knowledge (in Russell's phrase, "knowledge as love") is the attainment of a new and interesting way of conceptual seeing. But what about truth? It rings strangely on the ear to talk about "true" interpretations, though of course one wants to be able to distinguish between those interpretations that are adequate or right, and those that are not. Whewell wanted to be able to apply the predicate "true" as much as any philosopher of science. More so, since he wanted to account for the necessity of scientific conclusions. But the Galilean scientific realism to which Whewell aspired was ruled out from the start by his view of science as interpretation. This explains much that baffled his friends and foes alike, Herschel and DeMorgan as well as Mansel and Mill. Whewell talked about science enabling us to "see the facts in a new light," where the new way of seeing was expressed in a proposition that is necessarily true (note the inevitable conceptual slip). Such expressions of necessity cannot be comprehended by all men; training and long exposure to the ways of mathematics and science are required before one "sees the light." But at the same time Whewell did talk a lot about truth as if he meant it. He used terms like "true," "probable," "confirmed," as frequently and seemingly as innocently as any other philosopher of science. It is hard to see that the use was innocent. For Whewell had no theory of probability, no confirmation theory, and no correspondence model of truth. For probability theory he substituted the aesthetics of scientific success, for confirmation theory he substituted a novel theory of what it is that experiments actually disclose, and for correspondence he substituted consistency. Though he talked about testing and the hard attempts to get hypotheses to fit the facts, he usually was more interested in pointing out that science benefits more from the use of hypotheses that are inaccurate and maybe even false.[4] Such is the perplexing opening move of our philosopher. Can there be any wonder in the resulting fact that he introduces not one, but two theories of scientific method?

III. A QUALIFICATION: DISCOVERY AS JUSTIFICATION

Before I discuss in some detail the six central concepts in Whewell's inductive logic, let me introduce one more item that adds to the confusion. With respect to Reichenbach's distinction between "context of discovery" and "context of justification," Whewell was both historically and spiritually a pre-Reichenbachian. DeMorgan chastised him for confusing the two contexts in his argument that Whewell was confused about what logic is. Whewell replied that his induction ought to be called "discoverer's induction," an admission that I take to be equivalent to a willingness to conflate the two contexts.[5] Those of us for whom the very granting of a Ph.D. depended upon our acceptance of Reichenbach's distinction may find Whewell's talk about "logic" and "discovery" in the same breath philosophically intolerable in the extreme. Yet it is not all that difficult to understand that Whewell viewed science as an historically developing process, the end results of which are not complete, and perhaps never will be. There is, however, one neglected and essential event that takes place again and again in the playing out of this process: scientists, for good or ill, rightly or wrongly, take some propositions to be established once and for all (by induction), or, as Whewell would more than likely express it, there are times when the confidence of a scientist in his results becomes certitude, becomes grounds, explicable or not, for acceptance of a theory.[6] Whewell clearly wanted his theory of method to capture this prized kind of historical event and to explain it. His desires may well be seen as distorting his sense of what logic is, his insistence on keeping science in historical context may well be seen as rendering inexplicable his attempt to show that induction is demonstrative, but in all the confusion something interesting emerges—his second theory of induction.

What Whewell needed, then, was a master theory that would accommodate both of his background strategies, the one resulting from his view that science is interpretation or explication, and the one resulting from his view that scientists do, at certain points in time, come to accept theoretical and experimental results as established once and for all. I turn now to the details of the master theory.

IV. COLLIGATION AND THE MUTUAL DEPENDENCE OF THEORIES AND OBSERVATION LANGUAGES

Whewell's exposition of the classical hypothetico–deductive theory of science is probably the most masterful one written before the philosophy of science became a full-bodied discipline in the twentieth century. His exposition has never been bettered in its richness of historical detail—Whewell's understanding of science was that of the historian and the working scientist, not primarily that of the philosopher. I note this fact about Whewell only because my own interests in discussing him are philosophical, and I am liable to present a rather abstract version of his thought, partly to limit discussion to manageable length, partly because I am interested in discovering the philosophically arresting turns of Whewell's thought. It may not be fair to omit the rich variety of historical example characteristic of Whewell's writings, but omit it I must.

For Whewell, science begins with a special conceptual act. Events that are to be explained unfortunately do not offer their own suggestions for explanation. We must impose conceptual form on the materials of experience. We superinduce an idea on the facts, we gather apparently unrelated facts together in a conceptual net, we "colligate the facts," as Whewell would have it.[7] The colligating act, however, is more than a recording of recognized similarity; it is a rudimentary form of generalizing as well. Each colligation would have to be of the form "all of these noticed x's are also y's," or "some of these noticed x's are y's," or some other suitably quantified expression. Interpretation of facts is there at the beginning; the very first conceptual act is one of generalization through interpretation. Of course not every colligation is correct or valid; some of the suggested generalizations turn out to be false. It is the task of inductive logic to show wherein consists validity in inductions, for only those colligations that are correct count as valid inductions.

Two points are worth noting about Whewell's concept of colligation, especially since, in the exchange with Mill, Whewell takes the superinduction of an idea on the facts to be the defining property of inductions. Read one way, there is little novelty in Whewell's notion that in an induction a new idea is superinduced on the facts. Indeed, this is precisely what Hume and others seem to have had in mind when they claimed that inductive inference is not deductive, that is, that inductive conclusions are not part of what is meant by the ideas ingredient in the premises of a given argument. In this sense, Whewell's seemingly novel claim turns out to be nothing more than his noticing an apparently recalcitrant feature of induction.[8]

There is, however, a much more revealing way of construing Whewell's concept of colligation via the imposition of an idea on the facts. This construal is also more in keeping with what I take to be the major strategies of Whewell's theory of science. On this reading, each proffered colligation becomes a candidate for inclusion in a (hopefully standard) observation language. Each superinduced idea thus claims a place as a predicate in the observation language of science. In a special sense, then, the observation language of science is, as many recent writers have claimed, theory-laden. For each conceptually generalizing move made in a colligation involves imposed concepts. Keep in mind that science does not try to catalogue, but tries to explain. If we read Whewell's insight about colligation in this way, we see at once certain important consequences and problems. One important consequence is that decisions about which observation language to adopt are theoretical and also empirical decisions. The decisions are theoretical in that concepts are already involved. Only when theories of certain kinds are acceptable will the predicates generated by those theories be taken as legitimate observables. The decisions are empirical in just the sense that only when colligations are inductions, valid for the range of phenomena for which they supply explanations, will they be taken as supplying predicates that can function in something like a standard observation language. In this respect—and I think Whewell was one of the first to see this point—theories and observation languages grow up as surgically inseparable Siamese twins. What counts as an observable is that which true theories talk about; what counts as a true theory is that which talks about observables resulting from valid inductions.[9]

That this view yields problems will surprise no one. Whewell's view of colligation, construed in the second way, seems to lock theory construction and theory confirmation in a dark and impenetrable room. And the room revolves, enclosing a familiar kind of philosophical circle. If the theory introduces just the ideas (predicates) that explain the data, and the data, so to speak, cannot speak for themselves, how can a theory ever turn out to be incorrect? Put another way: how can a theory, given that it is required for any colligation of facts, ever turn out to be wrong, incorrect, or false? Will it not be the case that every explanation is as good as any other—that every colligation will be valid—on the quite sufficient grounds that only theory-imposition constitutes induction? Whewell recognized that such problems arise from his way of thinking about colligation, and his inductive logic was supposed to solve them. At the same time, throughout his career he held on to the view that theory and fact are ontologically indistinguishable (though of course they can be distinguished conceptually). The basic identity of theory and fact in

Whewell's methodology has an important bearing on his view of confirmation and experimentation. I will return to these matters below.

V. "TESTS" OF HYPOTHESES: PREDICTION

Whewell next moves to detail what he calls the various "tests" of hypotheses (generalizations originating from colligation) in order to show how one distinguishes between valid and invalid colligations. The entire script reads like a set-up for the hypothetico–deductive model, with some fascinating, and frequently unnoticed, asides. The hypotheses will already possess some degree of generality, and, in the best cases, will be formulated as mathematical equations. Deductive entailments of the colligations can be sought, and can be taken, when found, as confirming cases of the hypothesis in question. The first level of "test" of a hypothesis results from successful prediction, where the hypothesis explains additional facts *of the same kind* as those from which our rule was collected."[10] Successful predictions give us some confidence that our hypotheses accord well with what must be nature's laws; indeed the correspondence between untried facts and the entailments of a hypothesis "implies some large portion of truth in the principles on which the reasoning is founded."[11] Thus, "The prediction of results, even of the same kind as those which have been observed, in new cases, is a proof of real success in our inductive processes."[12]

Notice the hypothetico–deductive structure thus far articulated: hypotheses must account for what has been already observed to happen. Being general in form, hypotheses must also account for all other particular occurrences of the same kind as those already observed. We thus have two levels of epistemological support, each deductively involved in the other. More important than this, however, is the fact that Whewell uses two different vocabularies in talking about such epistemological links. He talks about explanation at both levels as "proof," "evidence," and "verification," and he speaks of the "truth" of hypotheses thus evidenced. In the alternative language, he speaks of the "success" of our inductive moves, and of the "conviction" that we are right in our suppositions. Most commentators (including the present writer in earlier works) have supposed that this second language is no more than a matter of psychologistic style, replaceable in all cases by the more sacrosanct epistemic phrases of the first language. I now think that the differences between the two languages should be taken seriously as an indiction that Whewell had more in mind than the familiar hypothetico—deductive model for scientific method. In the end it may turn out that we should always replace the epistemic locutions with terms from the second language.

Whewell realized, however, that success of predictions is still a systematically local matter. This test of theories does not help us to explicate those bold hypotheses that finally organize an entire field of apparently disparate data, maximum-information-seeking hypotheses like the inverse-square law of Newton. There is a difference between explaining all motions of a certain kind, and explaining all motions. So we arrive at the next level of tests of theories, a level at which our inductive moves enable us to explain facts of "a *kind different* from those which were contemplated in the formation of our hypothesis."[13] These are facts "unforeseen and uncontemplated"[14] when we formulate the hypothesis, but, though unexpected, such facts may from time to time turn out to be deductive consequences of the hypothesis in question.[15] Whewell calls the realization of this phenomenon a "consilience of inductions," thus introducing the most novel concept in his theory of method.

VI. "TESTS" OF HYPOTHESES: CONSILIENCE

A consilience of inductions takes place when a hypothesis introduced to cover one class of facts is later seen to explain another, different class of facts. This happens when two or more hypotheses are found to be deductive consequences of another, much more general hypothesis. Apparently the consilience takes place where a more general theory explains the data covered by two different hypotheses. More usually, it is two *laws* that become consilient by being derived from one more general theory. Thus Newton's theory of universal gravitation explained both the perturbations of the moon and the precession of the equinoxes, classes of phenomena thought originally to be quite dissimilar. What is more important than this rudimentary definition of "consilience" (a more detailed one will be given below), however, is what Whewell thought that we can claim epistemologically for such occurrences. Successful prediction contains some measure of truth; consilience impresses "us with a conviction that the truth of our hypothesis is certain."[16] It looks very much as though the concept of consilience becomes for Whewell the basic concept of his logic of induction. For if we can achieve justified certainty in some of our inductions, what harder test of inductive validity would we need? I think it is true that the concept of consilience is the touchstone of Whewell's logic of induction, but much more needs to be said before we can evaluate fairly his claim that in cases of consilience, induction yields certainty. From this point on, consilience is bound up with each important criterion for evaluation of theories—simplicity, successive generalization, induction as demonstrative—that Whewell introduces. A more detailed account of consilience must therefore be attempted.

The concept of consilience is an explanatory notion. Consilience is achieved when, among other things, enormous increases in the deductive–entailment content of a theory occur. It is imprecise to account for consilience by saying simply that different lower-level hypotheses become contained in a more general theory; indeed, it may be incorrect to say that in all cases such deductive containment holds (although Whewell speaks as though this will always be the case). Whewell wants well articulated scientific theories to be deductively connected throughout (for purposes of the logic of induction), but there is nevertheless a prominent sense in which consilience "has a history," that is, consilience is not an atemporal test of theories, it is a *happening*. (The conflation of discovery and justification is nowhere more clearly seen in Whewell than at this point.) I propose the following formal model of Whewell's concept of consilience:

Given two evidence classes E_1 and E_2, and two laws L_1 and L_2, at time T_1, L_1 explains E_1, and L_2 explains E_2, and there are no inductive reasons for supposing that L_1 and L_2 are connected. Put differently, at T_1 it is *thought that* E_1 and E_2 are disjoint.

Now at time T_2 a theory T is introduced. L_1 and L_2 become consilient at T_2 with respect to T when they both become logically derivable from T, that is, when T explains (because it entails) both L_1 and L_2, and, derivatively, explains E_1 and E_2. At T_2, then, E_1 and E_2 are no longer thought to be disjoint. (The same model holds, of course, if L_1 and L_2 are laws offered in explanation of other laws.)[17]

Given consilience in this sense, the evidence for, say, L_1 is increased because now part of the domain of facts covered by L_2 can be said to be explained by L_1. It is in this respect that Whewell takes consilience to be a better test of hypotheses than simple prediction. There are problems with this concept of consilience as additional confirmation of theories. At T_1 the two evidence classes E_1 and E_2 were thought to be disjoint because, on Whewell's account, they would have to be, for the simple reason that both L_1 and L_2 introduce new and clearly different ideas superinduced on the two different classes of data. Given Whewell's definition of induction, L_1 and L_2 have to be logically independent; otherwise they are identical. Thus, for consilience to occur at T_2 something more than simple deduction of L_1 and L_2 from T has to happen. Despite appearances, and many of Whewell's own pronouncements to the contrary, consilience cannot be explicated, specifically on Whewell's own grounds, as increase in confirmation through merely deductive connections between the laws and the theory involved.[18] The reason for this is simple: the theory T would itself be built up by induction (in Whewell's sense), and thus would involve the superimposition of a new idea on some data (not necessarily, at least originally, either E_1 or E_2). Science is interpretation; T involves an introduced interpretation; thus L_1 and L_2 can only become consilient with respect to T because T alters the meaning of some of the key terms of both laws, where such alteration renders the laws non-independent because they now share some reinterpreted terms in common relative to T. At the historically later time T_2, T deductively entails L_1 and L_2, only because basic terms of L_1 and L_2, and perhaps also some of their background conditions, have been altered in light of T. Some might even want to say that the observational bases of L_1 and L_2 have been altered by the introduction of T, thus also altering key terms used in describing the events described by E_1 and E_2.[19] It seems to follow that Whewell's concept of consilience cannot be explicated on purely deductive lines.

On the one hand, he insists upon deductive connections in all parts of sophisticated scientific theories; he even insists, as I shall detail below, that one of the defining characteristics of induction is that inductively derived propositions be deductively connected. This insistence yields his form of the hypothetico–deductive model of scientific method. On the other hand, his own theory requires that for the introduction and preservation of such deductive connections, severe changes in the meanings of key theoretical and observational terms take place. It is difficult to see, given this second consideration, how he can take either prediction or consilience as straightforward tests of hypotheses, since both "tests" appear to be almost self-guaranteeing as to truth, given the severe adjustments in a total theory that the inductive act of introducing new ideas requires. It would seem, then, that something more than direct deductive connection is required by Whewell for deciding between theories. Theories with clear deductive connections between all sentences ingredient in those theories, theories with great entailment content, are not hard to find—they can, as Whewell realized, be invented in great profusion. However it appears to the contrary, Whewell's suggested "tests" of hypotheses cannot be construed in simple confirmation–deductive explanation terms; extra-evidential considerations always seem to be both relevant and required in the assessment of theories.

The problem now confronting Whewell's system is the one of making good on the claim that consilience is a test of truth, or stronger, a test of certainty.[20] He was fond of pointing out that in the history of science no theories that have contained consilient hypotheses have later turned out to be false.[21] This historical claim may or may not be true, but even if it is true, this fact does not justify consilience as a test of truth. However persuasive it is, Whewell's historical observation fails to do the required epistemological job. He tries to do better; indeed, I think he gives two answers to the problem of consilience as a test of truth, one stemming from his straightforward hypothetico–deductive account in *Novum Organon Renovatum,* the other suggested by the novel features he introduces into this same account, plus some additional elements introduced elsewhere in his writings.[22]

Whewell held that the progress of science involved a process that he called "successive generalization."[23] No phrase could have been better chosen to bring out the hypothetico–deductive side of his methodology. In this view, science progresses by becoming more general; science seeks more and more powerful, more and more deductively inclusive, generalizations. Whewell invented a special device, called an "Inductive Table,"[24] for dramatizing this view of science. The tables show that the elementary facts of a given science are linked, via intermediate hypotheses and laws, to an all-inclusive unifying theory. Reading from the facts to the theory, one gets the order of inductive discovery; reading from the theory to the facts, one gets deductive justification of the whole array of hypotheses and laws in the form of maximal deductive content and hence of maximal deductive explanation. Such tables exhibit, throughout, the successive generalization character of science; they also amply display occurrences of consilience and the tendency toward greater and greater simplicity characteristic of well articulated sciences like astronomy. I have argued that on Whewell's own theory of induction the required deductive links can only be gotten by tampering with the semantics of the lower-level laws. Thus, though one can force a set of otherwise independent laws into deductive form, the resulting scheme (and the mirroring of this scheme in the inductive table) is contrived and deceptive in that it does not bring out clearly enough Whewell's own point that in each induction "the facts are seen in a new light," that is, that in each induction the higher-level law reinterprets other laws, gives other laws a different semantics, so that those laws now "exemplify," rather than confirm, the higher-level law.

VII. WHEWELL'S HYPOTHETICO–INDUCTIVISM

Whewell ought to have turned his attention directly to the question of what it is that justifies this semantical tampering; instead, he merely hints at an answer to this question[25] and goes on to give his hypothetico–deductive account of justification of induction. The account is perfectly standard except for the introduction of the contrivances called inductive tables. Science seeks ever more inclusive deductive form; the task of inductive logic is to provide a scheme whereby theories in deductive form can be seen to be arrived at validly via induction. Inductive logic, thought Whewell, is demonstrative, and ought to bring out the fact that valid inductive conclusions are necessary truths. He thus refers to his tables as "the criterion of truth"[26] of the laws they tabulate, and points out that the ultimate formula of inductive validity would have to be: "The several Facts are exactly expressed as one Fact *if, and only if,* we adopt the Conception and the Assertion of the inductive inference."[27] But his logic of induction gives no guarantee that we will ever arrive at justified use of this formula, though he remarks that "in reality, the conviction of the sound inductive reasoner does reach to [the] point"[28] of certitude. Whereupon he abandons the critical question of inductive logic with the capitulatory remark, "We may leave it to be thought, without insisting upon saying it, that in such cases what *can* be true, *is* true."[29]

The capitulation seems especially strange because Whewell does try something like a philosophical justification of his tables by stressing the deductive character of total inductive systems. What he seems to be after is some notion of explanatory tightness, some notion that will insure that the parts fit one another. He writes:

Deduction is a necessary part of Induction. Deduction justifies by calculation what Induction had happily guessed. . . . Every step of Induction must be confirmed by rigorous deductive reasoning, followed into such detail as the nature and complexity of the relation . . . render requisite. If not so justified by the supposed discoverer, it is *not* Induction.[30]

Nowhere in Whewell's writings is there clearer evidence for those who wish to construe his methodology in strictly hypothetico–deductive terms.[31] One might well leave the matter there[32] and credit Whewell with the historical merit of having presented one of the clearest statements of this form of methodology. Such a move, however, leaves out of account the distinctive features of Whewell's second—and I think for him more important—methodology.

VIII. CONSILIENCE, SIMPLICITY, AND SUCCESSIVE GENERALIZATION

As should have begun to appear, what I am calling "Whewell's second theory of method" emerges side by side with his more orthodox statement of the hypothetico–deductivist view of theories. For present purposes, I return to the model of consilience discussed earlier. Trying to be faithful to many of Whewell's statements, I deliberately formulated the model in deductivist terms. In turn, trying to be faithful to Whewell's own theory, I argued that the model must be recast as a model of semantical rearrangement of meanings of terms in lower-level laws. This shift in the model is important for a variety of reasons that must now be discussed in detail. In addition, if the differentiating points of the shift are not noted, I will repeat a mistake made in the Introduction to *William Whewell's Theory of Scientific Method* pointed out in the review of this book by Laurens Laudan, namely, the mistake of thinking that consilience increases confirmation by increasing content, that consilience is present where an acceptable maximum of content increase has been achieved by induction.[33] It is natural enough, working within the terms of Whewell's first theory of method, to think of consilience as an important measure of increase in entailment content of a theory. But Laudan's suggestion that I had oversimplified the concept of consilience led me to rethink the matter. I now think that consilience is a much more complex measure of theories than I had earlier thought.[34]

If, as I have shown, Whewell abandons the task of justifying induction as demonstrative (on any theory of demonstration current either then or now), it does not follow that he also abandons any attempt to show how we decide between theories, how we come to accept certain theories rather than others. It is in the area of this problem that Whewell's concept of consilience comes to play its important and perhaps irreplaceable role.

I will now argue for the thesis that Whewell's consilience, in his own special theoretical sense, is not a property of all inductions, certainly not a property of single conclusions arrived at by induction. Consilience is a property of some, and only some, well articulated (in the logical sense) theories which possess large measures of deductive content or which contain predicates that are expressively very rich. Consilience is thought to be a property of those systems having the following characteristics: 1) the theories must be simple (in a sense yet to be explicated); 2) the theories must be so general that they have almost reached the point of unity; 3) the theories must provide the best explanation of the large range of objects involved—for Whewell, to accept consilience as a test of inductive truth is to accept consilience arguments as arguments to the best explanatory scheme; and 4) the theories must have achieved that historical position where further testing of the law is seen to be irrelevant to acceptance of the theories, that is, the theories must have attained the position where negative results will be taken as calls for refinement of the systems, rather than as disconfirmations.

In other words, highly consilient theories are more acceptable than less consilient ones or ones exhibiting no consilience at all, because consilient theories display both systematic and extra-systematic features of those theories taken to be the best explanations of the data involved. Whewell expresses this point by regarding as identical the three most prominent features revealed by the inductive tables: consilience, simplicity, and successive generalization are concepts meaning the same thing.[35] Whewell holds this view because he takes important historical cases of consilience to be exemplifications also of simplicity and successive generalization. As he says, we should

direct our attention to two circumstances, which tend to prove, in a manner which we may term irresistible, the truth of the theories which they characterize:—the *Consilience of Inductions* from different and separate classes of facts;—and the progressive *Simplification of the Theory* as it is extended to new cases. These two Characters are, in fact, hardly different; they are exemplified by the same cases. For if these Inductions, collected from one class of facts, supply an unexpected explanation of a new class, which is the case first spoken of [consilience], there will be no need for new machinery in the hypothesis to apply it to the newly-contemplated facts; and thus, we have a case in which the system does not become more complex when its application is extended to a wider field, which was the character of true theory in its second aspect [simplicity]. The Consiliences of our Inductions give rise to a constant Convergence of our Theory towards Simplicity and Unity [successive generalization].[36]

This passage is extraordinary, and has not, to my knowledge, been noted by any of Whewell's commentators. If consilience were a measure of the entailment content of a theory, then any high-content theory, including those containing many *ad hoc* hypotheses, and those whose generality was not unified, but dispersed over many different classes of data, would be acceptable. On the other hand, if consilience were a measure of corroboration, then any theory which has withstood repeated severe tests, and which contains cases of consilience in the simple sense of deductively entailed laws that are logically independent, would have to be deemed acceptable, because not falsified; whereas Whewell wants to call such theories "true," and to take consilience as generating criteria for full acceptance of theories, given the realization, in individual cases, of the scientific goal of maximally successful explanation.

In the terms of the passage cited, two laws become consilient when they become deductive ingredients in a theory whose predicates are powerful enough to explain both of the different classes of data over which the laws separately ranged. No new *ad hoc* hypotheses will be required; but what will be required is a semantical reinterpretation of the terms in which the laws were earlier stated. Strictly speaking, the laws as originally formulated are eliminated in favor of laws whose predicates are semantically richer via interpretation given by the theory. Thus we begin to see how crucial it is for Whewell to insist that in *every* induction a new idea is superimposed on the data. If repeated hard tests or great increase in entailment content were all that were needed to prove inductions, then neither consilience nor simplicity would be possible, in Whewell's full sense of those terms. A hypothesis could never range beyond its original data without the imposition, in the semantics of a theory, of a new idea that could make it so range. Thus Whewell's insistence that in an induction the facts are seen in a new light is absolutely basic to his concept of consilience. He seems to be saying the following.

Each induction allows us to see facts in a new light. At some point, namely when "enough" of the data (for an experienced, gifted scientist one datum could be sufficient) in the original class are seen in this light, we can see those data in no other way. It is as if he were saying that each single induction over a domain of data is self-validating, in its own terms. But if induction were thus limited to completely valid single cases, no explication of the generality and explanatory power of full scientific theories would ever be forthcoming. Separate classes of data explained by logically independent laws could never be seen to connect either partially or completely, and scientific laws would have to be viewed as names on a list, as telephone book entries. Thus every induction from laws is undertaken with certain aims in mind; the search for new ideas is not haphazard nor purposeless, though it may be undertaken by the method of trial and error. Science becomes a quest for just those ideas that will relate separate classes of data, that will successively generalize in the direction of greatest simplicity. And if corroboration or content increase are not relevant tests of the truth of theories, then truth itself becomes equivalent to simplicity, and coherence of the parts of a system along certain lines becomes the test of truth. Scientific systems are acceptable when they achieve a certain form; the task is to find a theory that will give them that form.[37]

IX. INDUCTION AND THE ROLE OF EXPERIMENT

It is interesting that though his writings abound in references to "tests" and the role of "experience" in science, Whewell nowhere discusses the nature of experimentation in any detail. The reason for this is partly bound up with his full theory of necessary truth, a matter into which I will not enter here.[38] The reason is also connected with his second theory of method: if testing (in any usual sense of experiment) is irrelevant at the last stages of evaluation of theories, there is not much need for detailed treatment of it. On the other hand, Whewell does talk constantly about rigorous comparison of hypotheses with the facts, and at least some attempt must be made to understand what he means; all the more so, since the role of empirical considerations in evaluation of theories is in the end eliminable. There are two places in his writings where Whewell talks about the role of experiments in science (apart from what can be gotten by implication from his theory of necessary truth). In surveying these texts, we will find that Whewell has nothing like a theory of confirmation, but that what he does say greatly illuminates his move to the second theory of method.

In an early paper read before the Cambridge Philosophical Society,[39] Whewell provided an acute analysis of the nature of the truth-claims made by Newton's laws of motion. His problems were partly historical, partly philosophical. Historically, the laws had come to be regarded as necessary truths; philosophically the question of their warrant to be regarded as necessary must be considered, given that they are contingent claims about the world. In an ingenious set of analytic moves, Whewell separates the *a priori* (necessary) parts of the laws from their empirical parts, concluding that the form of the laws is given in thought, and that the empirical part of each law amounts in each case to the denial of the proposition that the conditions of a moving body are in any way causes of changes in circumstance of the motions of these bodies. The denial, of course, implies that all changes in motion of bodies are initiated by external forces. Now it is this proposition, this denial, that must be subjected to empirical test. But Whewell's move at this point is novel, and cannot be understood, I think, except in the context of his second theory of method. Instead of giving details of the kinds of tests that would be relevant in physics (hopefully generalizing this to apply to all cases of testing), he invokes the concept of simplicity in a curious way in order to show that *actual testing is irrelevant*.

He begins his argument by noting that the laws of motion are the simplest possible, since "they consist in the negation of all causes of change, except those which are essential to our *conception* of such causation."[40] Other conceptions are of course possible. We might, for example, take motions of bodies to be dependent upon lapses of time, or upon the motions that bodies have prior to being interfered with, or upon forces that have previously acted upon them. But none of these conceptions mirror reality; we do not find that we have to add more and more complex explanations in order to understand motion. He concludes:

The laws which, in reality, govern motion are the fewest and simplest possible, because all are excluded, except those which *the very nature of laws of motion necessarily implies*. The prerogative of simplicity is possessed by the actual laws of the universe, in the highest perfection which is imaginable or possible. *Instead of having to take into account all the circumstances of the moving bodies, we find that we have only to reject all these circumstances. Instead of having to combine empirical with necessary laws, we learn empirically that the necessary laws are entirely sufficient.*[41]

Further, Whewell adds that "all that we can learn from experience is, that she has nothing to teach us concerning the laws of motion. . . ."[42] and "The laws may be considered as a formula derived from *a priori* reasonings, where experience assigns the value of the terms which enter into the formula."[43]

Notice that Whewell does not here explicitly refer to inductive consilience, nor to simplicity in the full sense of simplicity of theories. But surely his reference to the simplicity of the laws of motion is no accidental remark. Whewellian testing is not a form of seeking for instances of a hypothesis that either confirm or refute it—indeed, his account says nothing at all about the possibility of finding the laws to be false. Consider his analysis of the necessary and empirical parts of the first law. The necessary part of the first law is summed up in the statement "Velocity does not change without a cause" (call this N), and the empirical part in the statement "the time for which a body has already been in motion is not a cause of change of velocity" (call this E).[44] One might think, put in this way, that empirical testing of the law would be designed to prove the empirical component E. Not so for Whewell. For the regular reoccurrence of E in specific cases is taken to mean that N is regularly and constantly exemplified by specific empirical cases, namely, that N as interpreted to mean that all causes of velocity will be external forces is always exemplified, via the regular and constant occurrence of events described by E. Earlier in his analysis Whewell claimed that N is a consequence of the completely *a priori* axiom that "Every change is produced by a cause." This axiom does not seem to prejudice any particular empirical outcome. Acceptance of this axiom would be quite consistent with accepting the negation of E. But the axiom, when applied to generate N, construed as a commitment to external forces, is incompatible with not-E.

Fixed up in this way, it is easy to see that E's truth tells us nothing that we did not already know, tells us that we do not have to account for other possible causes (in this case, lapse of time). Whewell's *a priori* causal axiom thus guarantees the acceptability of N and E, with no tests being necessary. It is in this sense that a Whewellian test of the empirical component of a law of motion is no test in the ordinary sense at all, because experience cannot fail to exemplify the combination of N and E on the prior interpretation of the axiom as covering only those causes that are external forces. If facts are never seen for what they are in their naked selves, if understanding of facts always presupposes imposed concepts, then there is some sense in trying to replace theory of confirmation by theory of exemplification. If science is the right interpretation of the facts, the *facts* cannot determine the rightness of the interpretation by themselves. Rather, it seems to be the case that we want something that we demand of *theory* to decide between competing alternative explanations.[45] Whewell wanted consilience and simplicity to decide these issues, he wanted certain elements in a structure of rational assessment of theories to count more heavily.

X. CONFIRMATION OR EXEMPLIFICATION

I want to pause over this suggested replacement of confirmation by exemplification, if only because much recent ink has been spilled on it by some philosophers of science. Whitehead put it succinctly: "experiment is nothing else than a mode of cooking the facts for the sake of exemplifying the law."[46] Feyerabend has also put the point (in discussing the meaning of Newton's "deriving the laws from the phenomena"):

Newton's "phenomena," which are the elements of the new "experience," are not everyday facts pure and simple; nor are they an experience that has been cleared from prejudicial elements and left that way. They are rather an intimate *synthesis of laws,* possessing instances in the domain of the senses and certain mathematical ideas. . . . Actual experiment, which always depends on a large variety of irrelevant variables, may therefore *illustrate* the phenomenon; it cannot *establish* it. . . .[47]

Whewell's treatment of experiment in his discussion of the laws of motion seems to me to put him in the Whitehead–Feyerabend camp. Indeed, Whewell's account provides a good example of how categorial theory-ladenness of observations prejudices the result of an experiment beforehand. But look a little more closely at this curious concept of "exemplification" or "illustration."

It might seem at first glance that confirmation and exemplification are symmetrical notions. For, after all, does not an exemplification of a law also confirm it? Does not a confirmation of a law also exemplify it? Surely for a law to be either confirmed or exemplified it must be instanced, but its instances seem quite indifferent to whether we take them to confirm or to exemplify. But this has to do with positive cases only. Consider the questions: does a failure to illustrate a law disconfirm it, and does a disconfirmation of a law fail to illustrate it? Again it might seem that failure of either process entails failure of the other. The processes of confirmation-disconfirmation and exemplification-non-exemplification appear to be quite symmetrical, if not actually identical. I think, however, that this apparent symmetry is an illusion. Consider the following (greatly oversimplified) schemata:

(1) An experimental result *e confirms* a law L, if and only if a description of *e* is a deductive expectation in a system containing L, and the description of *e* is true.

(1′) An experimental result *e exemplifies* a law L, if and only if a description of *e* is a deductive expectation in a system containing L, and the description of *e* is true.

The schemata, given their identity, bring out again the suggestion that confirmation and exemplification are on all fours when it comes to putting the positive cases. But something like a difference emerges when we put the negative cases. For (1) has a meaningful opposite in not-(1), which is the familiar case of disconfirmation. The question is: does (1′) really have a negation? From the point of view of exemplification, can not-*e* ever occur?

I think the answer to these questions is negative. To say that a law is exemplified is just to say that it has instances in the domain which it was introduced to explain. In Whewell's terminology, the law will be exemplified by just those kinds of things which the law's new concept picked out in the first place. Thus any occurrence of a not-*e* is impossible, for negative cases are not the kind of thing that the law explains. Anything that looked like a not-*e* would thus be eliminated as experimentally irrelevant. (Note again the way in which Whewell rigged the laws of motion in order to eliminate the non-exemplifying cases beforehand.) But suppose that quite a large number of apparent not-*e*'s appeared as experimental results? In this case, Whewell seems to have thought, we will modify the law, or refine the law, but we are not logically obliged to give it up. For a non-Whewellian, the concept of exemplification seems to be contextual; it is all a question of how a law or hypothesis is entertained. If we take the hypothesis in question to be suspicious and needing further evidence, we talk of confirmation. If we take it to be established (in whatever methodological sense of establishment), we then talk only about modifying or refining it. However, for Whewell, exemplification is not a contextual concept in this sense, because for him the appropriate context is always the same, given that the new concept introduced in an induction can never fail to be exemplified. Put another way, there will never be any logical integrity in the negative cases. If such cases appear to occur, they will not force abandonment of the law, though they might be taken to force theoretical adjustments in the law or in the system in which the law is ingredient.[48]

Whewell also discusses the question of what we may expect from experimental results while analyzing Newton's rules of philosophizing.[49] His general point about the rules is that the philosophy of science that underlies them is inductivist, and that the rules thus need reconstrual in terms of Whewell's own theory of science. On two specific counts he indicts Newton's understanding of his own rules. Count one is that rule 3 (the qualities of bodies which cannot be increased or diminished in intensity, and which belong to all bodies in which we can institute experiments, are to be held for qualities of all bodies whatever) gives too much weight to the authority of experience, since experience cannot establish the universality of any properties of bodies whatsoever. Count two is that rule 4 (in experimental philosophy, propositions collected from phenomena by induction are to be held as true either accurately or approximately, notwithstanding contrary hypotheses, till other phenomena occur by which they may be rendered either more accurate or liable to exception) is mistakenly taken by Newtonians to mean that inductive propositions may from time to time be given up because of the logical force of negative experimental cases.[50] That Whewell wins his case for the first count seems to me to go without saying. On the other hand, Whewell's reading of Newton's fourth rule needs much argument (and I am suggesting that his second theory of method provides that argument). Inductivist readings of rule 4 take Newton to have meant that there is a point at which inductive conclusions must be given up, and that that point is reached when massive disconfirmation has been achieved after repeated tests. Whewell, with his *a priorist* confidence in the conclusions of inductive arguments, advances the position that the comparison of a hypothesis with others will always have nugatory results, given that the hypothesis has been established by induction in the first place. Induction being (in some sense) demonstrative, laws established by induction can be refined on the basis of apparently recalcitrant experience, and on the basis of this same experience we can list exceptions to the laws, but *those laws cannot be falsified*. All of which again strongly argues the point that for Whewell, Popperian corroboration plays no role whatsoever in scientific induction.

XI. A SUMMARY OF WHEWELL'S SECOND THEORY

I will not labor here points about Whewell's theory of method that I have made elsewhere. Instead, I want now to gather together the results of the present investigation. I have been arguing that though Whewell does indeed present one eloquent form of the classical hypothetico–deductive method of induction, there are too many novel features of his own account that do not at all fit that model. Starting with his axiom that science is the interpretation of the meaning of nature, one can only conclude that the major elements of his inductive logic—induction as imposition of new ideas on data, consilience, simplicity, and successive generalization—argue for a different theory of science. Additional buttressing of this conclusion comes from his way of treating the question of the role of experimental results. One cannot ignore the fact that he stresses the importance of deductive connections between statements listed in his inductive tables, but one would misread him badly if he counted only this part of Whewell's theory. His own novel suggestions might have led him to put his inductive logic in quite different terms. Instead of stressing the deductive character of general theories, he might have made more of (and I have been arguing that he did make more of) his theory that the inductive tables express the fact that choices between theories are made on the basis of certain criteria of success. Instead of successive generalization, Whewell might have spoken of "successive exemplification," given the fact that the inductive tables were designed to show that a certain "act of attention" is required to see that at each inductive stage, the new idea involved did indeed inductively express the data or laws below it on the table.

That the history of inductive science should be read by Whewell as a structure of successive exemplification is important, given his notion that the distinction between facts and theories is only relative, or does not exist at all. Beginning with the most humble empirical generalization, whose terms are exemplified by "data," and arriving at unified theories exemplified by less general laws, science exhibits a closed structure that Whewell labels "the idealization of facts."[51] We can of course imagine a great variety of alternative idealization schemes. The problem is how to choose between them. In the end, for Whewell, a valid inductive argument is an argument to the most consilient scheme, namely, to that idealization scheme that best organizes, through reinterpretation of key terms, the lower-level hypotheses and "data." If my way of reading Whewell's second theory of induction is right, then we can concede Whewell's historical analysis of the steps of discovery without much argument. The inductive tables schematize that history; they do not, whatever Whewell claims to the contrary, schematize a list of valid inductions in ways analogous to deductive validity schemata. Further, the schematization exhibits the various *forced* stages of exemplification of higher laws, where the force is exerted by new theories at higher levels. A corollary of these exemplification stages is a mutual adjustment of observation language and theory as a given scientific system enlarges and explains more. Finally, the schematization of the inductive tables displays cases of consilience, simplicity, and successive generalization. The major point of the second methodology of Whewell is that consilience and its correlates be taken as marks of the success of a certain scientific system in achieving maximum explanatory power. His point is partly historical—each level of discovery marked out on the tables signifies a time at which in fact scientists did think themselves successful in achieving a new explanation—and partly a claim about how theories are to be assessed—the assessment always depending upon features of scientific theories that relate in relevant ways to the aims scientists have actually stated as the aims of their inquiry.

Whewell's second theory of method is perhaps unsatisfactory. Successive exemplification leads us in the end to those root concepts whose ultimate justification can only be given in theological terms.[52] On Whewell's view science leads us out of Plato's cave. Philosophical dissatisfaction with this model should not, however, blind us to the fact that Whewell's theory of science is remarkably faithful to one clear historically obvious feature of the scientific enterprise—basically, science seeks explanation. Although one cannot, in ordinary inductivist ways, determine the factual truth of scientific theories, this revelation ought not to keep us from seeing that there are other "tests" of scientific systems. These tests, as Whewell has shown in his second theory, are applied to decisions about the success of scientific systems in achieving what scientists want. Whewell was one of the first to see clearly the goal-directed character of science. The insights that this clear vision made possible enrich our understanding of the complexities of actual science, and point out the direction that inductive logic should take in future.

REFERENCES

Agassi, Joseph. 1969. "Sir John Herschel's Philosophy of Success." In *Historical Studies in the Physical Sciences*. Vol. 1, pp. 1–36. Edited by Russell McCormmach. Philadelphia: University of Pennsylvania Press.

Butts, Robert E. 1965a. "Necessary Truth in Whewell's Theory of Science." *American Philosophical Quarterly* 2:1–21.

———. 1965b. "On Walsh's Reading of Whewell's View of Necessity." *Philosophy of Science* 32:175–81.

———. 1967. "Professor Marcucci on Whewell's Idealism." *Philosophy of Science* 34:175–83.

———, ed. 1968. *William Whewell's Theory of Scientific Method*. Pittsburgh: University of Pittsburgh Press.

———. 1970. "Whewell on Newton's Rules of Philosophizing." In Butts and Davis, pp. 132–49.

Butts, R. E. and J. W. Davis, eds. 1970. *The Methodological Heritage of Newton*. Toronto: University of Toronto Press.

DeMorgan, Augustus. 1859. Review of Whewell's *Novum Organon Renovatum*. *The Athenaeum* no. 1682.

————. 1860. Review of Whewell's *Philosophy of Discovery*. *The Athenaeum* no. 1694.

Ducasse, Curt J. 1960. "John F. W. Herschel's Methods of Experimental Inquiry." In Madden, pp. 153–82.

————. 1960. "William Whewell's Philosophy of Scientific Discovery." In Madden, pp. 183–217.

Feyerabend, Paul K. 1965. "Problems of Empiricism." In *Beyond the Edge of Certainty*, pp. 145–260. Edited by R. G. Colodny. Englewood Cliffs: Prentice-Hall.

————. 1970. "Classical Empiricism." In Butts and Davis, pp. 150–70.

Herschel, Sir John. 1830. *Preliminary Discourse on the Study of Natural Philosophy*. Reissued with an introduction by Michael Partridge. New York: Johnson Reprint Co., 1966.

————. 1841. Review of Whewell's *History* and *Philosophy of the Inductive Sciences*. *Quarterly Review* no. 135, pp. 142–256.

Hesse, Mary. 1968. "Consilience of Inductions." In Lakatos, pp. 232–46.

————. 1971. "Whewell's Consilience of Inductions and Predictions." *The Monist* 55:3, pp. 520–524.

Lakatos, Imre, ed. 1968. *The Problem of Inductive Logic*. Amsterdam: North Holland Publishing Co., pp. 232–57.

Laudan, Laurens. 1970. Review of Robert E. Butts, ed. *William Whewell's Theory of Scientific Method*. *British Journal for the Philosophy of Science* 21:3, pp. 311–312.

————. 1971a. "William Whewell on the Consilience of Inductions." *The Monist* 55:3, pp. 368–391.

————. 1971b. "Reply to Mary Hesse." *The Monist* 55:3, p. 525.

Mackie, J. L. 1968. "A Simple Model of Consilience." In Lakatos, pp. 250–53.

Madden, Edward H. 1960. *Theories of Scientific Method: the Renaissance through the Nineteenth Century*. Seattle: University of Washington Press.

Marcucci, Silvestro. 1963. *L' "Idealismo" Scientifico di William Whewell*. Pisa: Istituto di filosofia.

————. 1969. *Henry L. Mansel*, pp. 298–301. Firenze: P. Le Monnier.

Mill, John Stuart. 1874. *A System of Logic*. 8th ed. New York: Harper & Brothers.

Walsh, Harold T. 1962a. "Whewell and Mill on Induction." *Philosophy of Science* 29:3, 279–284.

————. 1962b. "Whewell on Necessity." *Philosophy of Science* 29:2, 139–145.

Whewell, William. 1858. *History of Scientific Ideas*. London: John W. Parker and Sons.

————. 1858. *Novum Organon Renovatum*. London: John W. Parker and Sons.

————. 1848. "Second Memoir on the Fundamental Antithesis of Philosophy." *Transactions of the Cambridge Philosophical Society* 8: 614–20.

Whitehead, A. N. 1955. *Adventures of Ideas.* New York: New American Library.

NOTES

1. Whewell's work is finally receiving much-needed attention. See Butts 1965a, 1965b, 1967, 1968, and paper in Butts and Davis 1970; Ducasse in Madden 1960, 183–217; Walsh 1962a, 1962b. Herschel's methodological work is still largely neglected, though a new issue of his *Preliminary Discourse on the Study of Natural Philosophy* in 1966 may help. See Ducasse in Madden 1960, 153–82; and Agassi 1969, 1–36.

2. Whewell, *Novum Organon Renovatum,* Aphorism I, "MAN is the Interpreter of Nature, Science the right interpretation"; and Whewell, *History of Scientific Ideas,* "The course of real knowledge is, to obtain from thought and experience the right interpretation of our general terms, the real import of our maxims, the true generalizations which our abstractions involve" (268).

3. Whewell's Platonism is discussed by Marcucci 1963 and 1969, 298–301; and by Butts 1967.

4. Whewell, *Novum Organon Renovatum,* in Butts 1968, 149–151. All references to Whewell's works, unless otherwise specified, will be to selections in Butts 1968.

5. DeMorgan 1859, 1860. The exchange is discussed in the introduction to Butts 1968, 24–26.

6. Butts 1968, 173.

7. Butts 1968, 138–177.

8. Even on this interpretation of colligation there remains some novelty and interest, however. For Whewell's objection to Mill's claim that we see the ellipse in the data (the discussion had to do with Kepler's discovery of the elliptical paths of the planets) can be read as suggesting that if this is all there was to Kepler's discovery, then either he simply observed the ellipse, or the ellipsoid nature of the paths was an idea already ingredient in earlier parts of Kepler's theory, in which case the elliptical path laws are deductive consequences of other parts of that theory. But observation and deduction are not induction. Thus, on any sensible reading of Mill's claim that the ellipse was in the data, he was quite wrong in having thought that Kepler performed an induction. Such considerations seem to have convinced Whewell that he was right. In

fairness to Mill, however, it should be mentioned that he thought that Kepler's induction was an argument to the conclusion that the paths will remain elliptical. Whewell seems not to have understood the nature of such generalizing arguments, at least in the context of his exchange with Mill. See Whewell, "Mr. Mill's Logic," Butts 1968, 272–277; and John Stuart Mill, *A System of Logic,* bk. III, ch. II, sections 2–4.

9. Whewell expresses this point in language quite different from that used in this paper. In "On the Fundamental Antithesis of Philosophy" (Butts 1968, 54–75), he argues for the ultimate indistinguishability of theories and facts. In *Novum Organon Renovatum* (Butts 1968, 176–77), he argues that the distinction is only relative and that it is in any case untenable.

The following passage from "On the Fundamental Antithesis" makes my point about the connectedness of theory and observation language:

In the progress of science, both the elements of our knowledge are constantly expanded and augmented. By the exercise of observation and experiment, we have a perpetual accumulation of facts, the materials of knowledge, the objective element. By thought and discussion, we have a perpetual development of man's ideas going on: theories are framed, the materials of knowledge are shaped into form; the subjective element is evolved; and *by the necessary coincidence* of the objective and subjective elements, the matter and the form, the theory and the facts, each of these processes furthers and corrects each other: each element moulds and unfolds the other. (Butts 1968, 75)

Whewell makes the same point by suggesting a form of *entrenchment* as a characteristic of the building up of observation languages:

Theory and Fact are the elements which correspond to our Ideas and our Senses. The Facts are facts so far as the Ideas have been combined with the sensations and *absorbed in them:* the Theories are Theories so far as the Ideas are kept distinct from the sensations, and so far as it is considered as still a question whether they can be made to agree with them. A true Theory is a fact, a Fact is a *familiar* theory. (Butts 1968, 59, emphasis added)

10. Butts 1968, 153.

11. Ibid., 152.

12. Ibid.

13. Ibid., 153.

14. Ibid.

15. Of course the results in a consilience are unexpected; they have to be, given that an inductive idea maps out completely the domain that it will cover. But no ordinary deductive, or probabilistic, explication of consilience evidence can capture the sense in which consilience makes a law more acceptable. I hope to show this more clearly below.

16. Butts 1968, 153.

17. A similar model has been provided by J. L. Mackie, "A Simple Model of Consilience," in Lakatos 1968, 250–251.

18. Mary Hesse's "Consilience of Inductions," in Lakatos 1968, 232–246, is extremely helpful in displaying what I take to be the typically Whewellian point that "a confirmation theory can only explicate consilience of inductions if the language that is built into it is the language of the relevant scientific theory" (239). On a somewhat different path, Mackie holds that "we get confirmation by consilience only where the consilient inductions *exemplify* a single principle or theoretical law" (Mackie, 252). I have found both of these papers enormously helpful in trying to get a good grip on Whewell's concept of consilience.

19. Surely Whewell would want to say this, given his position on the coincidence of theory and data outlined in note 9 above. Whewell's writings abound with references to semantic and conceptual change, although, unhappily, a great many of his discussions contain more metaphor than analysis. In the wonderful essay, "Of the Transformation of Hypotheses in the History of Science" (Butts 1968, 251–262), he discusses the question of how rival explanations in science get to be decided, concluding:

And thus, when different and rival explanations of the same phenomena are held, till one of them, though long defended by ingenious men, is at last driven out of the field by the pressure of facts, the defeated hypothesis is transformed before it is extinguished. Before it has disappeared, it has been modified so as to have all palpable falsities squeezed out of it, and subsidiary provisions added, in order to reconcile it with the phenomena. It has, in short, been penetrated, infiltrated, and metamorphosed by the surrounding medium of truth, before the merely arbitrary and erroneous residuum has been finally ejected out of the body of permanent and certain knowledge. (262)

Whewell's stress upon the imposition of a concept in cases of induction (and hence of consilience) serves to underscore his view of progress in science as involving semantic changes in scientific theories. For example, in his discussion of the inductive table of optics, he regards the undulatory theory of light as having won the field, largely through successive consiliences achieved by the use of the theoretical concept of polarization. But on his own admission, polarization achieves consiliences of the various phenomena of light only on the supposition that undulations are transverse (Butts 1968, 157). That vibrations are transverse is not a newly discovered datum, it is introduced as a new and partial meaning of the term "undulation," and, thus understood, polarization in terms of the undulatory theory of light accommodates all of the phenomena of light. Whewell is quite prepared to generalize from such examples. Thus he writes:

In Induction . . . , besides mere collection of particulars, there is always a *new conception,* a principle of connexion and unity, supplied by the mind, and superinduced upon the particulars. There is not merely a juxta-position of materials, *by which the new proposition contains all that its component parts contained; but also a formative act exerted by the understanding, so that these materials are contained in a new shape. . . .* Our Inductive Tables, although they represent the elements and the order of these Inductive steps, do not fully represent the whole signification of the process in each case. (Butts 1968, 163, emphasis added)

In addition, Whewell himself admits that cases of induction, and hence also cases of consilience, are not fully explicable in terms of inclusion of less general propositions in more general ones.

But when we say that the more general proposition *includes* the several more particular ones, we must recollect what has before been said, that these particulars form the general truth, not by being merely enumerated and added together, but by being seen *in a new light.* (Butts 1968, 169–70)

20. Butts 1968, 173–76.

21. For example, see Butts 1968, 154–55.

22. Whewell's theory of induction is much more complex and rich than he himself often took it to be. Driven by an almost Hegelian delight in unified theories, he tended to read the history of science as a quest for deductive structure and deductive unity. The novel features of his theory of method do not rest well on the procrustean bed of this ancient doctrine. Whewell's own realization of the tensions within his system seems to have come in debate rather than in exposition. Thus in the exchanges with both DeMorgan and Mill he appears willing to abandon questions of deductive form in favor of stressing the new insights of his theory, e.g., colligation of facts as imposition of a new concept, inductive acceptance of total theories on the basis of extraevidential principles like simplicity.

23. Butts 1968, 160–77.

24. Ibid.

25. When, in the discussion of the inductive tables, he refers to the special "act of attention" required to see that the facts actually do fit the hypothesis (Ibid., 168–69), and when, in discussion of Newton's rules of philosophizing, he distinguishes rather clearly between the process of confirmation and that of refinement of a law already taken to be true (Ibid., 333–36; also see my paper in Butts and Davis 1970, 143–47). I will return to these points below.

26. Butts 1968, 176.

27. Ibid., 174.

28. Ibid., 173.

29. Ibid.

30. Ibid., 175-76.

31. I have discussed some of the novel non-hypothetico-deductive features of Whewell's discussion of the inductive tables in the Introduc-

tion to Butts, 1968, 18–24. Since my aims in this paper are somewhat different from those in the Introduction, I will not repeat those remarks here.

32. Ducasse seems to have done so in Madden 1960, 216–17.

33. Laurens Laudan, 1970, review of Butts 1968.

34. Though I thus concede Laudan's point that consilience is not always or uniquely bound up with content increase in theories, I will not concede what I take to be the implications of his remark that the Popperian notion of severe tests must be considered in explicating Whewell's notion of consilience. Consilience is not achieved when increase in content is achieved; neither is it achieved when increase in *corroboration* has been achieved. What I say below about Whewell's concept of an experiment may help to clarify the points of this new disagreement with Laudan.

While writing this paper I read Laudan 1971a and b and Hesse 1971 in typescript. I am much indebted to both authors (and to Laudan for acute private discussion of the issues involved in consilience). In a recent lecture delivered in the University of Pittsburgh Series in the Philosophy of Science ("Consilience of Inductions and the Problem of Conceptual Change in Science," October, 1971), I argued that Hesse's formalization of the concept of consilience is preferable to Laudan's, and that the concept of consilience adds nothing to our ordinary understanding of the probabilistic confirmation of laws or theories. The new papers by Laudan and Hesse do not seem to force any important changes in the interpretation of Whewell's inductive logic offered in the present essay.

35. Butts 1968, 159–60.

36. Ibid., 159.

37. Additional evidence for my way of reading what I am calling Whewell's second theory of scientific method may be gotten from Whewell's way of interpreting Newton's first rule of philosophizing: "We are not to admit other causes of natural things than such as both are true, and suffice for explaining their phenomena." Briefly, Whewell collapses the distinction between "true cause" and "adequate explanation," arguing that direct ontological criteria be replaced by an ontological criterion having to do with the success of theories. Both consilience and simplicity play a role in this argument. Again, briefly, scientific entities are said to exist (to be "true causes") when hypothetical constructs naming them are ingredients in successful scientific theories, where those theories are powerful in explanation, i.e., exhibit consilience and simplicity. I have discussed these matters in Butts and Davis 1970, 139–42.

38. For detailed discussion of Whewell's concept of necessary truth, see Walsh, 1962b; and Butts 1965a, 1965b.

39. "On the Nature of the Truth of the Laws of Motion," in Butts

1968, 79–100. Additional discussion of this paper is in Butts 1965a, 9–13.

40. Ibid., 98, italics added.

41. Ibid., italics added.

42. Ibid., 99.

43. Ibid., 100.

44. Ibid., 97.

45. Sir John Herschel, in his early review of Whewell's *History* and *Philosophy* (Herschel, 1841), noticed this feature of Whewell's theory of experimentation. He wrote:

Experience, according to [Whewell], only exemplifies, cannot prove a general proposition. Its truth stands on the higher and independent ground of *inherent necessity,* and is recognized to do so by the mind so soon as it becomes thoroughly familiarized with the terms of its expression. (173)

Mackie (see note 18 above) also picks out this exemplification feature of Whewell's theory of science.

46. *Adventures of Ideas,* 94.

47. P. K. Feyerabend 1965, 159–60; see also his discussion in Butts and Davis 1970; and application to Whewell in Butts and Davis 1970, 140–46.

48. In this respect, Whewell appears to have also hit upon what is now called Duhem's thesis about the non-falsifiability of physical laws.

49. In Butts 1968, 333–36.

50. I discuss this in detail in Butts and Davis 1970, 143–47.

51. Whewell, 1848, "Second Memoir on the Fundmental Antithesis of Philosophy," 33–35.

52. Butts 1965a, 13–19.

4 Logic of Discovery in Maxwell's Electromagnetic Theory[*]

MARY HESSE

University of Cambridge

I

Contemporary accounts of the structure of science are broadly distinguishable into two types which I shall call the hypothetical and the historicist.[1] The hypothetical accounts restrict the operation of logic and reasoning to the deductive testing of hypotheses and banish the role of discovery to psychology or mere guesswork. The historicist accounts on the other hand describe the internal coherence of particular theories, emphasizing how the presuppositions of empirical observation itself are theory-laden and historically relative. Neither account investigates the inductive relation of observation to theory; indeed, both in a sense deny that there can be any such logical relation, the first because the only logic available is that of deduction from hypotheses, and the second because observation is already permeated by theory and therefore cannot be used as independent inductive evidence for theory.

Nevertheless, the historian of science cannot ignore the number of occasions on which inductive evidence for theories has been claimed, and inductive reasoning found acceptable. This feature of scientific inference has no doubt been oversimplified in the past, and therefore in order to investigate its character and validity it has become neces-

* I am indebted to Dr. S. d'Agostino, Mr. J. Dorling, and Dr. P. M. Heimann for helpful comments on a first draft of this paper, and also to the participants in the discussion at the conference on "Foundations of Scientific Method" at Indiana University.

sary to re-examine the historical cases where inductive inferences are claimed. It is likely that as a result of such re-examination the traditional accounts of both inductive and deductive inference in science will be found to be inadequate.

In this paper I shall investigate in particular Maxwell's explicit discussions of physical method, and their application in his electromagnetic theory. This theory has given rise to much historical literature, in which almost every style of scientific inference recognized by philosophers has been claimed to be illustrated. But in spite of this it has not been much remarked that Maxwell himself often claims that his method is the authentic Newtonian one of "deduction from experiments" without the aid of unproved hypotheses. My aim is to clarify this claim and to consider how far it is justified in his electromagnetic theory, particularly in the introduction of the electric displacement current, which looks on the face of it like a paradigm case of the "theoretical concept" of later philosophical analysis. I shall argue that, on the contrary, in his mature theory Maxwell intended it to involve no hypothetical concept, and no physical "model" as ordinarily understood, but that he attempted to justify it by a generalized method of induction and analogy.

In Maxwell's discursive methodological remarks[2] four different types of theoretical method can be distinguished: the hypothetical, the mathematical, the analogical, and "deduction from experiments."

1. The Hypothetical Method

In several investigations Maxwell explicitly adopts the classic hypothetical and eliminative method. In his early work on Saturn's rings, for example, he lists three possible mechanical hypotheses as to their form: they are solid and uniform, or not solid, or not uniform. A solid uniform ring would be unstable; the form of solid irregular ring which would be stable is inconsistent with observation; hence the hypothesis of a fluid ring remains to be investigated. And in the first paper on the dynamic theory of gases, the hypothesis of small, hard, perfectly elastic spheres is to be explored:

If the properties of such a system of bodies are found to correspond to those of gases, an important physical analogy will be established, which may lead to more accurate knowledge of the properties of matter. If

experiments on gases are inconsistent with the hypothesis of these propositions, then our theory . . . is proved to be incapable of explaining the
phenomena of gases. In either case it is necessary to follow out the consequences of the hypothesis. (*Scientific Papers,* 1:378)

In both these examples it is noticeable that the hypotheses are
entirely mechanical in character, that is to say, no physical quantities
are involved except those concerning masses in motion and the forces
of weight, impact, pressure, and friction. When the subject matter
is less obviously mechanical, however, Maxwell is more critical of
the hypothetical method. In electric and magnetic science, he says,
if "we adopt a physical hypothesis, we see the phenomena only
through a medium, and are liable to that blindness to facts and
rashness in assumption which a partial explanation encourages"
(1:155). Theoretical entities such as molecules and the aether
should not be postulated without evidence (2:253, 315), and should
not be endowed with *ad hoc* "attractive and repulsive forces when-
ever a new phenomenon is to be explained" (2:339, cf. 223). And
in the case of chemistry, where the material systems are too small to
be directly observed, the hypothetical method is only amenable to
verification "so long . . . as someone else does not invent another
hypothesis which agrees still better with the phenomena" (2:419).
Not only was the hypothetical method regarded by Maxwell as un-
desirably pervasive in nineteenth-century physics, but his objections
to it rest on arguments, not on timidity or prejudice.

2. The Mathematical Method

On the other hand Maxwell is not content with theories composed
of purely mathematical formulae in which "we entirely lose sight of
the phenomena to be explained; and though we may trace out the
consequences of given laws, we can never obtain more extended
views of the connexions of the subject" (1:155). In a review of
Thomson's *Papers on Electrostatics and Magnetism* he complains
that no one has developed Thomson's theory of vortex molecules:
"Has the multiplication of symbols put a stop to the development of
ideas?" (2:307), and in a paper of the same period he writes:

We must retranslate [symbols] into the language of dynamics. In this
way our words will call up the mental image, not of certain operations

of the calculus, but of certain characteristics of the motions of bodies. (2:308)

Explicit criticism of the purely mathematical method does not go very deep in Maxwell's writings, but it has been pointed out interestingly by G. E. Davie that dislike of pure analysis is ingrained in the Scottish tradition in which Maxwell had his first philosophical education.[3] Maclaurin's method of geometrizing the calculus was not dead in Edinburgh, and Maxwell's own interest in physical interpretations of Euclidean geometry is exhibited in several of his early mathematical papers. We shall see later, however, that logical as well as genetic reasons for objecting to purely formalist methods in physics emerge implicitly in Maxwell's treatment of physical theory.

3. The Analogical Method

For Maxwell the middle way between the "rash assumptions" of physical hypotheses and the "analytical subtleties" of mathematical formulae consisted sometimes in a method of "physical analogy," and sometimes in a Newtonian method of deduction of forces from phenomenal motions. The first of these methods appears in the introduction to his first paper on electricity and magnetism, "On Faraday's Lines of Force" (FL).[4] Faced with the dilemma between hypothesis and pure mathematics,

We must therefore discover some method of investigation which allows the mind at every step to lay hold of a clear physical conception, without being committed to any theory founded on the physical science from which that conception is borrowed. . . .
In order to obtain physical ideas without adopting a physical theory we must make ourselves familiar with the existence of physical analogies. By a physical analogy I mean that partial similarity between the laws of one science and those of another which makes each of them illustrate the other. (1:156)

Maxwell here gives four examples, which recur throughout his subsequent discussions of physical analogy: the laws of numbers on which all mathematical sciences are founded, the resemblance of form between both corpuscular theory and wave theory and the phenomena of light, William Thomson's analogy between electric and magnetic attraction and the equations of heat conduction and fluid flow, and the analogy which Maxwell develops in *FL,* between theories of electric and magnetic action at a distance, fluid flow, and Faraday's representation in terms of lines of force.

The precise nature and function of this kind of "physical analogy" are not altogether easy to gather from Maxwell's explicit remarks about them, and I shall consider them in more detail below. As a preliminary statement it may be said that on the one hand Maxwell is concerned to insist that the existence of a "formal" analogy of equations does not imply identity of physical process or substance, as when fluid flow is compared with heat flow, with current, and with electric induction, without any implication that these last processes in fact involve fluids in motion. The function of formal analogies is rather to aid the imagination in understanding formal relationships, and to enable transfer of mathematical results from one system to another irrespective of subject matter. But on the other hand, systems of ideas which are "really analogous in form" must be distinguished from those that are merely mathematical, and when such analogies are found, they lead "to a knowledge of both [systems], more profound than could be obtained by studying each system separately" (2:219), and "It becomes an important philosophical question to determine in what degree the applicability of the old ideas to the new subject may be taken as evidence that the new phenomena are physically similar to the old" (2:227).[5] Again, although causes of invisible processes cannot be identified with their formal analogues in observable processes, the relations between cause and effect are similar, and what we require are methods of representation so general that they express the real similarity of relations without introducing unwarranted hypothetical ideas into the expression of the cause.[6]

4. Deduction From Experiments

These general methods of representation are to be "deduced" from experiments by a method similar to that used by Newton in deriving the law of gravitational force. Echoing Newton's own statement of method, Maxwell contrasts "the true method of physical reasoning [which] is to begin with the phenomena and to deduce the forces from them by direct application of the equations of motion," with the "too frequent practice" of inventing a particular dynamic hypothesis and deducing results, the agreement of which with phenomena "has been supposed to furnish a certain amount of evidence in favour of the hypothesis" (2:309).[7] In electrical science, however, the motions involved are not all observable, and the forces involve quantities which are not mechanical:

When we pass from astronomical to electrical science, we can still observe the configuration and motion of electrified bodies, and thence, following the strict Newtonian path, deduce the forces with which they act on each other; but these forces are found to depend on what we call electricity. (2:419)

For hypothetico–deductivist philosophers of science, talk of "deduction from experiments" must appear logically bizarre. I do not wish here to examine the logical credentials of the expression, which has been common since Newton, except to remark that of course no one has claimed that general laws can be deduced from particular experiments without intervention of a general premise of some kind. For Newton one premise was supplied by some version of his "Rule of Philosophizing III," to the effect that "The qualities of bodies . . . which are found to belong to all bodies within the reach of our experiments, are to be esteemed the universal qualities of all bodies whatsoever."[8] For Maxwell one such premise was always the conservation of energy, supplemented in his later writings by the Lagrangean formulation of the laws of mechanics. These premises were themselves regarded, however, as generalizations from experience, powerful enough to enable specific laws to be deduced for systems other than those from which they were induced.[9] Maxwell developed the method explicitly in the *Treatise,* where he proposes

to examine the consequences of the assumption that the phenomena of the electric current are those of a moving system, the motion being communicated from one part of the system to another by forces, the nature and laws of which we do not yet even attempt to define, because we can eliminate these forces from the equations of motion by the method given by Lagrange for any connected system. . . . I propose to deduce the main structure of the theory of electricity from a dynamical hypothesis of this kind.[10]

This preliminary account of Maxwell's analogical and deductive methods leaves their distinctive character obscure. For it may be argued that the method of physical analogy is nothing but the mathematical method in disguise, since many of Maxwell's remarks about it may be interpreted in the light of later formalist analyses of science as implying no more than similarity of abstract mathematical relations, in which no real or physical similarity between the analogous systems need exist. On the other hand it may be held that the method of deduction from experiments is essentially the hypothetical method, since it admittedly depends on general premises which are logically indistinguishable from hypotheses. I think both these arguments are mistaken, and for the same reason. For in both cases it is assumed that elementary inductive generalizations do not constitute a distinctive form of scientific inference which is distinguishable from hypothetical inference and also a necessary element in the recognition of physical analogies. I shall try to substantiate this claim, first by examining what Maxwell means by "real physical analogy," and second by showing that when such analogies are established, hypothetical theories and concepts may be avoided.

II

In later analyses of physical theory, formal analogy has sometimes been construed in terms of uninterpreted symbols related by equations (the "calculus"), for which entirely independent interpretations are given (by "correspondence rules") to transform the calculus into the analogous theories of different physical systems. If the formal analogy between two systems implied no more than this, it would be a rather weak if not trivial relation, for in the first place, the mathematical expression of the theory of a given physical system is generally not unique. It is therefore possible that in some forms of representation a formal analogy would exist between two systems, but not in other forms. In the second place, in the formalist view the theoretical calculus gives in itself no hints as to its own interpretation, and any interpretation by a set of correspondence rules which will make a given system a semantic model of a given calculus would be acceptable as a theory of that system. Thus, not only may a given system be represented by many formal theories, but a given formal theory may have many models in physical systems, some of them mutually contradictory, as in the case of the particle and wave theories of light. It follows from both these types of non-uniqueness that there can be no analogical argument from one system to another with which it has some formal analogy. For it is not possible to distinguish the "real" formal analogy from others produced by transformations of mathematical representation, or indeed to know whether any one of them can be said to be "real," and even if one such analogy is distinguished, there are no grounds for picking particular models of the formal theory as the intended ones in the analogous systems, and arguing from the known properties of one system to further properties of the other systems.

If Maxwell's formal analogy is not that of later formalists, the question arises, what constitute for Maxwell the constraints required to ensure that transformation from one analogous system to another represents real physical analogy and not merely arbitrary similarity of form? Some light on this question is thrown by what he calls the "mathematical classification" of physical quantities (2:257). Classification into scalars, vectors, and quaternions is an important example, as is the distinction among vectors between forces, represented by magnitude along a line, and fluxes, represented by the magnitude of an area, directed normally to that area. This classification of forces and fluxes conveys the mathematical distinction between many pairs of physical quantities: length and area, temperature and heat, electric potential and current. There is also a distinction among vectors between those referring to translation and those to rotation, as in the grad (Maxwell's "slope"), div (Maxwell's "convergence"), and curl functions. These functions represent linear and vortex fluid motion respectively, and also, by analogy, electric and magnetic actions. It follows that the mathematical entities involved—scalars, vectors, forces, fluxes, translations, rotations—are not uninterpreted symbols, but have at least a spatio-temporal interpretation which is identical in all physical systems to which they apply. The "flux" of current may not be flow of a substance, but it has direction in space and quantity related to cross-sectional area, and in these respects is exactly the same as the flow of water along a tube. Translation and rotation have equally obvious spatial interpretation in all Maxwell's analogous systems; indeed, in spite of his many disclaimers that the relata of an analogy need not themselves be at all similar, he assigns "rotation" to magnetic force precisely because it causes spatial rotation of the plane of polarized light, detected by physically turning polarization gratings in space. But what it is that causes the turning is (at least in his later papers) not specified in the case of magnetic force as it is in the case of material fluid vortex motion.

Transformation rules between formally analogous systems do not, however, always depend on identities of spatio-temporal properties. Sometimes they take the form of identifications of dissimilar causes which nevertheless have identical effects. The most obvious example is the identification of all forms of force with mechanical force, whether produced by a mechanical or electrical or magnetic or any other physical system. All forms of force satisfy the dynamic laws of motion, and hence have the same mechanica effects. They are therefore generically identical properties of systems. A less obvious example in Maxwell's electrical theory is the identification of the potential function of electrostatic force with the "electric tension" (potential difference) causing currents in conductors. This identification deserves scrutiny, because although physically elementary, it reveals a fundamental logical structure which has been little noticed in descriptions of scientific inference.

In *FL* Maxwell introduces the theory of current electricity by reference to the researches of Ohm and Kirchhoff, which had shown that in maintaining a current, "pressure" must be different at different points of the circuit:

This pressure, which is commonly called electrical tension, is found to be physically identical with the *potential* in statical electricity, and thus we have the means of connecting the two sets of phenomena. If we knew what amount of electricity, measured statically, passes along that current which we assume as our unit of current, then the connexion of electricity of tension with current electricity would be completed. . . . Thus the analogy between statical electricity and fluid motion turns out more perfect than we might have supposed, for there the induction goes on by conduction just as in current electricity, but the quantity conducted is insensible owing to the great resistance of the dielectrics. (1:180)

The argument is this: electric tension is the analogue of static potential and also of fluid pressure, in the sense that the experimental laws satisfied by these three systems are of the same form. Moreover, electric tension and static potential are physically identical. By this identity Maxwell seems to mean that, given a difference of potential between two points, its physical effect depends solely on the type of material existing between the points: if it is an insulator the phenomena are those of electric induction, if it is a conductor, those of electric current.[11] In this passage in *FL* Maxwell is inclined to carry the analogy further and suggest that it indicates also identity of the unobservable processes involved, namely that both current and electric induction consist of the flow of a substance. We have here the first hint that electric induction should be treated as physically similar to current flow, a step which leads eventually to postulation of the displacement current. But it is already clear that this extension of the analogy is a shaky one, for between the whole process of induction and of current there are obvious negative analogies: conductors and insulators have very different electric properties; in particular, current generates heat, and induction so far as we know does not do so.

However, the identification of tension with potential does not depend on this further analogical extension, which is in fact abandoned in "A Dynamical Theory of the Electromagnetic Field" (*DT*):

In the case of electric currents, the force in action is not ordinary mechanical force, at least we are not as yet able to measure it as common force, but we call it electromotive force, and the body moved is not merely the electricity in the conductor, but something outside the conductor, and capable of being affected by other conductors in the neighbourhood carrying currents. (1:539)

And, as we shall see later, even the model of current as "something flowing" in the conductor is not insisted upon in later expressions of the theory.

We now have a method of recognizing "real analogy" as contrasted with mere mathematical similarity of equations. It is simply that if two apparently distinct properties of different systems turn out to be interchangeable in appropriately different physical contexts, they are to be considered the same property. I shall call this type of inference to identity *experimental identification.* There may be *generic identification,* when it is the properties of different systems that are identified, or *substantial identification,* when the same entity is found to be involved in apparently different systems. An example of substantial identification is Maxwell's conclusion in "On Physical Lines of Force" (*PL*), on the basis of the identity of numerical value of the velocities of transmission of transverse electromagnetic waves and of light, that the aetherial media of electromagnetism and light are one medium (1:492).

Another important example of generic identification is the assumption that all forms of energy are to be identified. This assumption is indeed essential for the method of "deduction" of generalized laws of motion from experiments. It does not presuppose that all energy is reducible to the kinetic and potential energy of matter in motion, any more than the language of forces and fluxes implies that heat or electric current are so reducible. It does imply that some general terms can be substituted in Lagrange's equations for the "kinetic" and "potential" energy of mechanical systems, in such a way that the equations are satisfied by diverse physical systems, and that the real analogues of kinetic and potential energy in these other systems are observably identifiable. The identifications are somewhat more complex than the equivalence of effects which led to the identification of electric induction and potential, for they depend on transformations of energy according to quantitative conservation rules. The generic identity of energy justifies Maxwell's claim in *DT* that, whereas all other phrases inherited from the vortex model of the earlier *PL* "in the present paper are to be considered as illustrative, not as explanatory," still, "In speaking of the Energy of the field, however, I wish to be understood literally" (1:564).

Analogical inference leading to experimental identifications goes very deep into the structure of science. Such inferences are not infallible—the identification of phlogiston in different chemical reactions, for example, turned out to be a mistake—but they are pervasive in all scientific theory, and may be supported by stronger or weaker inductive evidence. They are indeed involved even where no specific arguments for identification are adduced. Consider the identification of mechanical force due to magnets, and to electrified bodies, with mechanical force in general. No argument seems to be required to establish this, but it is not a logically necessary identification. It rests on the interchangeability of mechanical effects of different kinds of physical systems, and the fact that "force" in all cases satisfies the laws of motion. The possibility of such identifications rests on an assumption of relative independence among physical properties, that is to say, although the logical possibilities of variety in different physical systems is unlimited, it is not supposed that interdependence within physical systems is in fact so tight that all the properties of every system have to be considered as different from all of those of every other system.

Similar considerations regarding identity throw some light on the nature of the "experiments" which form the starting point of Maxwell's (and Newton's) "deductions." In *DT* Maxwell claims to deduce his theory from Ampère's and Faraday's laws. These laws rest on such experimental identifications as that of all forces with mechanical force, magnetic force produced by permanent magnets with that produced by iron and by currents, and currents produced by batteries, with current produced by friction or electric induction. The experimental laws are not only empirically vulnerable in the usual sense of being generalizations from instances, they are also vulnerable in being analogical generalizations over instances of different kinds, leading to experimental identifications of properties without which there could be no general descriptive language with which to carry on scientific inference. It may be noted parenthetically that the occurrence of these identifications makes the task of any logic of induction doubly difficult, for not only is there the usual problem of explicating evidence for inductive generalizations, but also the stock of basic distinguishable individuals and predicates in the language is continually changing as evidence for identifications accumulates or is found to be misleading.

We can now summarize the method which Maxwell claims is a preferable alternative to both the hypothetical and mathematical methods, and spell out exactly how it is distinguished from them. First it should be clear by now that the methods of analogy and of deduction from experiments are not separate methods but aspects of a single method. The view that this is not so is due to a confusion between the method of analogy and the postulation of a hypothetical physical model, such as the mechanical vortex aether model in Maxwell's early electromagnetic theory. This model may exhibit some formal analogy with electromagnetic phenomena, but it is quite clear that even in *PL* Maxwell thought that it failed to be a satisfactory theory: "I do not bring it forward as a mode of connexion existing in nature" (1:486). The method of physical analogy, however, is different. It begins with two or more existing physical systems which are related in two ways: 1) they satisfy formally analogous mathematical laws, and 2) there are sufficient experimental identifications of their entities and/or properties to constitute a real physical analogy between the systems, and permit analogical inference from one system to the other. Under these conditions it is possible, without specifying in detail the character of unobservable entities or causes, to use the general laws representing the formal analogy of the systems in deducing the particular form of the laws and their particular effects for particular systems. For example, in the case of Newton's theory, experimental identifications of the properties of various mechanical systems as constituting analogues related by the laws of motion enables the particular form of the law of gravitation to be deduced from the observable laws of planetary motion. Both experimental identifications and the general laws of motion are necessary here, in order to permit laws derived directly from observable forces and motions to be transformed into laws concerning forces-at-a-distance, and bodies whose masses cannot be discovered by the usual operational means of scales and spring balances.

Enough has been said to indicate why the method of analogy and deduction from experiments is not a purely formal mathematical method. But it may be asked whether it is not after all just a particular form of the hypothetical method. It can indeed be represented in terms of the hypothetico–deductive schema, but there are the following crucial differences between it and the kind of hypothetical speculations to which Maxwell objects.

1. The general laws constituting the formal analogy between systems are themselves derived by direct inductive generalization from experiments; in other words they are "experimental laws" rather than hypotheses.

2. No hypothetical concepts or entities are postulated, since there

is no specification of the detailed character of unobservable causes except in so far as this is justified by experimental identifications and analogical argument using the general laws.

3. A logic of induction to laws and analogy between systems replaces the "guess and test" method of the hypothetico–deductive account.

III

I shall now examine Maxwell's "Dynamical Theory" of 1864 as an application of the analogical method. In this paper Maxwell abandons the hypothetical vortex model, and claims that his conclusions, particularly the equation for the displacement current, are "deduced from experimental facts." It would be at least charitable to suppose that this claim is meant seriously, and that Maxwell now believes himself to have a valid and non-hypothetical inference from experiments to his electromagnetic equations, even though we might prefer to describe this inference as inductive or analogical rather than deductive.

Maxwell begins by arguing the substantial identity of the electromagnetic medium and the light aether, in contrast to *PL,* where this identity is held to follow from the postulates of the theory. The phenomena of light and heat, he says, give reason to believe "that there is an aetherial medium filling space and permeating bodies, capable of being set in motion and of transmitting that motion from one part to another" (1:528). This medium must carry energy, since heat and light radiation take time to traverse it, and the conservation of energy through time is assumed. It is also assumed that the energy is mechanical, in the sense that it is carried as the energy of motion of the material aether, and of its "elastic resilience":

We may therefore receive, as a datum derived from a branch of science independent of that with which we have to deal, the existence of a pervading medium, of small but real density, capable of being set in motion. . . . (1:528)

Later on in the paper, even this minimal specification of the material aether is withdrawn, and all that remains of the properties of the medium is whatever can be identified respectively with the kinetic and potential energy terms in Lagrange's generalized equations of motion.

The next step in *DT* is to consider the evidence from motion of conductors and dielectrics relative to a magnetic field. When a material body is moved across magnetic lines of force, whether by the motion of the body or change of the lines, its ends tend to become oppositely electrified, and it may even experience chemical decomposition. By the same inference that led to the identification of electric tension and potential, the electromotive force is defined as that single property which causes the observable polarization, whether in conductors, dielectrics, or electrolytes. Maxwell goes on to consider the action of the electromotive force in dielectrics:

(11) . . . when electromotive force acts on a dielectric it produces a state of polarization of its parts similar in distribution to the polarity of the parts of a mass of iron under the influence of a magnet, and like the magnetic polarization, capable of being described as a state in which every particle has its opposite poles in opposite conditions.* (1:531)

The footnote is a reference to the work of Faraday and Mossotti on polarization of material dielectrics.

On the face of it, Maxwell's argument in this section (11) is an appeal to the formal analogy of electric and magnetic polarization developed experimentally by Faraday, and mathematized by Poisson and Mossotti. After reading Faraday, Mossotti had realized that Poisson's theory of magnetic polarization could be directly applied to dielectrics by, as Maxwell put it, "merely translating it from the magnetic language into the electric, and from French into Italian" (2:258). But this formal analogy in itself is hardly what Maxwell would call a "real physical analogy," sufficient to justify inference to the displacement current; indeed, the known relations between electricity and magnetism constitute rather a negative than a positive analogy.

In the paragraph following the one just quoted, however, Maxwell sketches Mossotti's own representation of dielectrics, which involved more than the skeletal mathematical analogy with magnetic polarization. The physical content of Mossotti's hypothesis depended on another analogy, namely that between the behavior of dielectrics and conductors in an electric field. Consider a large parallel plate condenser in which the plates are separated by free space. Its capacity is defined as the ratio of charge on one of the plates to potential difference between the plates, and is constant for constant geometric configuration of the plates. If macroscopic uncharged conductors are now introduced between the plates they are found to be polarized by the electric field, and the net effect on the potential distribution is found to be a decrease in the potential difference between the plates for a given charge, or in other words, an increase in the capacity of the resulting condenser, the amount of the increase depending on the size and position of the small conductors. Now this increase in capacity is exactly what occurs when a material dielectric is inserted between the plates, the amount of increase being characteristic of the geometry and dielectric constant of the dielectric medium. Mossotti's hypothesis is that material dielectrics consist of microscopic conducting particles insulated from each other by space or aether, where the dielectric constant of each specific material depends on the unobservable geometric configuration of its microparticles. In this hypothesis there are properly speaking no material dielectrics; the only dielectric (insulator) that exists is aether. It therefore provides a simple, physically realizable macromodel for microscopic processes, and may be said to compare favorably in inductive status both with the formal analogy of magnetic and electric potential and with Maxwell's vortex model.

Maxwell does not mention the details of Mossotti's hypothesis in *DT*,[12] but he may have regarded it as implicitly strengthening his analogy between conducting currents and displacement currents in material dielectrics. It causes no difficulty so long as it is restricted to material dielectrics, but trouble arises as soon as, following Faraday, Maxwell begins to treat the aether itself as a dielectric. He is bound to take some such step, since his view is that energy processes are going on in the medium between conductors and magnets, whether this medium is free space or filled with material insulator. What more natural, then, than to identify the structure of the aether with that of material dielectrics, and to postulate a charge polarization of its parts under the influence of electromotive force by analogy with that of material dielectrics? The cavalier manner in which Maxwell passes in *DT* from "real" electric current and "real" dielectric polarization to displacement current and polarization in aether suggests that he has used this analogy quite uncritically, and regards it as needing no further support. After an interlude chiefly devoted to Ampère's and Faraday's investigations of the mechanical effects of closed currents, he moves straight on to the "General equations of the electromagnetic field," and immediately asserts, "Electrical displacement consists in the opposite electrification of the sides of a molecule or particle of a body which may or may not be accompanied with transmission through the body. . . . The variations of the electrical displacement must be added to the currents $p, q\ r$ to get the total motion of electricity" (1:554). Maxwell subsequently substitutes this "total motion" for the ordinary conduction current j in Ampère's experimental relation for closed circuits, so as to obtain (in our notation) curl $H = (4\pi/c)j + (1/c)\dot{D}$, and uses this equation, without further comment, to refer both to material dielectrics and to the aether (1:557).

After so much insistence on deduction from experiments, the brevity of this most crucial part of the argument comes as something of a shock, and not surprisingly, many critics found it unconvincing. It is worth noticing, however, that Maxwell might at this stage have supplemented the analogical argument by being more explicit about his interpretation of the nature of the displacement current. His unwillingness to specify the character of the cause here prevented him from examining the inductive force that might have been given to the argument by analogies that lay close at hand, as I shall now show.

It was soon noticed by Maxwell's successors that the implicit interpretations of electric charge and current in his work are highly ambiguous and sometimes confused.[13] Broadly speaking, two distinct interpretations have been recognized, which I shall call simply the first interpretation and the second interpretation. The first interpretation regards electric charge as an incompressible substance satisfying a fluid continuity equation, whose motion in conductors constitutes current and which acts at a distance; the second regards charge as an epiphenomenon of the polarization of the aetherial medium, which becomes detectable when lines of force meet interfaces between insulators and conductors. The first interpretation is related to the traditional action at a distance view of electric force;

the second is derived from Faraday's view of the primacy of the polarization of the medium. There is no doubt, as several commentators remarked, that Maxwell's own use of the term "displacement current" caused confusion at the outset, because it falsely suggests that he is consistently adopting some variety of the first interpretation.

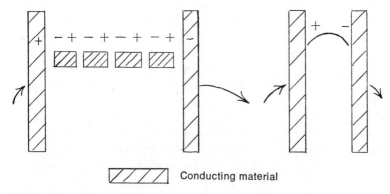

First Interpretation
with material dielectric
particles indicated

Second Interpretation
with no material dielectric
between plates

Certainly the *prime facie* interpretation to be given to Maxwell's words in paragraph (11) is the first. Unfortunately, this makes the argument depend on an analogy between material dielectrics and aether which becomes increasingly implausible the more it is scrutinized. For in Mossotti's hypothesis, the aether is just that which intervenes between the conducting particles of the material dielectric, and if particles carrying moving charges are now postulated of aether the question arises, what intervenes between the aether particles? At best the model reduces to one of action at short distances within the aether. There are admittedly places where Maxwell seems to allow this possibility, but it fits uneasily with his general insistence on the propagation of action through a medium.[14] A further physical difficulty of the first interpretation arises from the fact that charge is regarded as an incompressible fluid whose motion produces the total current of conduction and displacement, moving in closed circuits without sources or sinks.[15] In other words both current and electric induction are transmitted instantaneously, an assumption that is conceptually awkward in a theory of the electromagnetic aether in which wave-disturbances are found to travel with the velocity of light.

The first interpretation, then, does not satisfactorily bridge the gap between material dielectrics and aether in the argument to the displacement current, both because it rests on no close analogy between dielectrics and aether, and also because it leads in itself to implausible physical consequences. When we consider the second interpretation, however, in which charge depends upon lines of force, many of the difficulties of the first interpretation disappear, but at the cost of reducing the plausibility of another part of the analogical argument.

As we have seen, Maxwell argues in many places that energy of electric polarization resides in aether in some fashion which reveals itself to observation as a tendency to motion along the direction of the electric lines of force. On the second interpretation, the charge on a conductor is just the interruption of a line of force by a conducting surface, not some *thing* whose presence produces the lines of force. It is a misunderstanding to regard this as a circular definition of charge or to regard the first interpretation as somehow closer to observation than the second,[16] for both models are interpretations of what is observed. We do not first "observe" charges as particles, and then define lines of force; we observe motions of macroscopic bodies which have been treated in a certain way: by friction, by motion in the neighborhood of magnets, by being connected to a battery, and so on. All these observable situations are reinterpretable directly into the second interpretation without intervention of the idea of charged particles in motion. The cause of electric lines of force in aether is not the presence of substantial charges; the cause is friction, moving magnets, batteries, etc. Wherever a static or moving charge is mentioned in the first interpretation, it can be consistently reinterpreted as a state of the lines of force, at rest or in motion, in the second interpretation.

Again, the notions of incompressibility and conduction of charge acquire new interpretations. There is no motion of particulate or fluid charges along a conductor, and nothing corresponding to pressure along it either; therefore there is no problem about instantaneous transmission. When a condenser is charged, the second interpretation identifies this process with the establishment of a system of tubes of force between the condenser plates, and between these and other conductors if there are any present in the space. As the charging process takes place, the tubes of force move into position at right angles to their length, so that they become established at the surfaces of all terminating conductors simultaneously. There is no instantaneous action from one condenser plate to the other; it is an action of the battery or other charging device, in which energy takes time to reach the conductors from the battery through the field. There is also implicit in the second interpretation a radical shift in the notion of a current carrying conductor. This is no longer conceived on the model of material particles or fluids in pipes, but as a closed surface on the outside of which lines of force terminate, and along which they move, thus producing the effects of moving charge on the second interpretation. If this conception had had an observable model to replace fluids in pipes, the analogical argument to Mossotti's hypothesis, and hence to the displacement current, might even have been said to be complete.

It cannot be pretended that Maxwell conceived anything like a detailed picture of the second interpretation in *DT*. Its full consequences were indeed not presented until the papers of Poynting in 1884 and 1885 and Heaviside in 1885–1887, which showed that Maxwell's equations entail a continuous energy flow in the field where electric and magnetic actions are taking place.[17] In this they are only following out consistently Maxwell's original insight that energy is present in the field. In "On the Transfer of Energy in the Electromagnetic Field," Poynting writes:

If we believe in the continuity of the motion of energy, that is, if we believe that when it disappears at one point and reappears at another it must have passed through the intervening space, we are forced to conclude that the surrounding medium contains at least a part of the energy, and that it is capable of transferring it from point to point.

According to Maxwell's theory, currents consist essentially in a certain distribution of energy in and around a conductor, accompanied by transformation and consequent movement of energy through the field. (p. 343)

A conduction current then *may be said to consist* of this inward flow of energy with its accompanying magnetic and electromotive forces, and the transformation of the energy into heat within the conductor. (p. 351, my italics)

Poynting recognizes, however, that the inductive evidence for the theory is not increased by the new interpretation:

We can hardly hope, then, for any further proof of the law [of energy transfer] beyond its agreement with the experiments already known until some method is discovered of testing what goes on in the dielectric independently of the secondary circuit. (p. 361)

And in the *Note* written four years after *DT* Maxwell himself reacts to the hiatus in the experimental argument for the displacement current by requiring the same direct verification:

the current produced in discharging a condenser is a complete circuit, and might be traced within the dielectric itself by a galvanometer properly constructed. I am not aware that this has been done, so that this part of the theory, though apparently a natural consequence of the former, has not been verified by direct experiment. The experiment would certainly be a very delicate and difficult one. (2:139; cf. *Treatise on Electricity,* Section 607)

The same assessment of the logical situation was later made by Helmholtz and Hertz. Indeed, as the first page of his Introduction to *Electric Waves* makes clear, Hertz's own experiments were prompted by the offer of a prize by the Berlin Academy for experimental investigation of the relationship between electromagnetic forces and dielectric polarization which would complete just this deductive argument from experiment. In 1887 Hertz claimed to have shown the magnetic effects for some dielectrics, but despaired of verifying them directly for air. His eventual discovery of the finite propagation of electromagnetic waves in air is in fact derived from his earlier attempts at this direct verification for air.[18]

IV

It must be concluded that neither in the first nor in the second interpretation can Maxwell's analogical inference to the displacement current be made cogent.[19] A more fundamental question about his method remains to be asked, however, namely whether any such argument could be made to work in the light of objections stemming from the historicist or "meaning variance" account of science. The conclusion, "There is a displacement current in aether," the historicist will claim, means something quite different in the two interpretations; how then can it be said to be inferred without theoretical assumptions from the experimental facts? Consistently with Maxwell's discussion of formal analogy and deduction of generalized equations from experiments, the reply would seem to be this: the conclusion "There is a displacement current in aether," understood as asserting the occurrence of an extra term in Ampère's equation, does not depend for its experimental meaning or validity upon either interpretation of current. It is not an assertion of the nature of the unobservable cause of observable magnetic effects due to a changing electric field, it is a generalized expression of the laws relating these effects with other observables, namely the macroscopic set-up which constitutes the changing electric field. The displacement current has in this respect a status similar to that of the energy of aether, whose relations with observables are expressed by general laws of motion without specification of the kind of energy involved in mechanical or other microscopic terms.

   This reply leaves two further points to be clarified. First, if the assertion of the displacement current is intended to be independent of either interpretation, what force could the attempted analogical arguments have, even if they were successful? For the arguments certainly do depend upon adopting one or the other interpretation. It should be noticed, however, that if either or both of the analogical arguments had worked, the corresponding interpretation would not be objectionably hypothetical, because it would have rested on observable analogies of behavior of just the same kind as those leading to identifications of the forms of energy, and of electric induction and potential. There would also have been no need to regard the interpretations as alternatives with respect to the argument for the displacement current term in Ampère's equation, for an argument from one set of observed analogies is surely strengthened by an argument from another set just in so far as the arguments lead to the same conclusion, even though in other respects they may lead to different conclusions.[20] An inference to the sine law of refraction, for example, is stronger in virtue of being an inference in both the corpuscular and the wave theories.

The second point is more fundamental. The historicist has doubtless been waiting throughout the last two paragraphs to object that the whole force of his analysis has been missed by talking about the "observables" which are said to be related by the generalized "neutral" displacement current equation. These observables themselves, he will maintain, are pervaded by theoretical interpretation. Here there is, however, an ambiguity. If it is being asserted that there are theoretical assumptions hidden in descriptions of experimental observations, such as those concerning electric potential for example, this has already been taken account of. These theoretical assumptions have been specified above in terms of empirically vulnerable experimental identifications which underlie not only the experimental laws assumed, but also the descriptive predicates in terms of which they are expressed. In the latter case the experimental identifications may not rest on any explicit argument; they are directly recognized as acceptable uses of whatever descriptive vocabulary is currently in use. No undesirable regress is thereby created, because some empirically vulnerable identifications of "the same property again" must be presupposed in the use of any descriptive language. It does not follow, of course, that some such identification may not later have to be brought to consciousness, questioned, and possibly abandoned in the light of further evidence, as was the case with the tacit identification of electric current with "something flowing" (presupposed in the very metaphor of "current").

In order to describe the concept of displacement current neutrally as expressing a relation between observables, it must be assumed that in both interpretations of current the same experimental identifications are assumed in expressing the observables. It follows that if the historicist objector wishes to claim that no distinction can be made between observables which are neutral relative to the two interpretations and the interpretations themselves, he is in conflict with the present account. Without entering into this dispute in detail, it may be remarked that the historicist's mistake here lies in his assumption that the only alternative to "radical meaning variance" between theories is a radical distinction between theories on the one hand and an absolutely neutral observation language on the other. In the present account an intermediate position is adopted: theory is not to be conceived, as in the hypothetico–deductive account, as an external hypothesis *imposed,* however intimately, upon independent observations; it is rather to be understood as *constituted by* the fundamental experimental identifications which control both the inference to general laws, and the basic descriptive language in which the laws are expressed. In this, the present account agrees with the historicist. But it does not follow in this account that all theories are self-contained and incommensurable, as some historicists have maintained, for some experimental identifications span more than one theory, and can be used as the relatively neutral ground of comparison between theories.

In summary, then, I have tried to show how Maxwell's method of analogy and deduction from experiments can be developed in such a way as to provide an alternative, not only to the hypothetical and mathematical accounts from which he himself wished to distinguish it, but also to the historicist account which would deny both the concept of deduction from experiments, and also, more fundamentally, the independent status of experiments themselves.

NOTES

1. The distinction is familiar in recent literature: the hypothetical (or hypothetico-deductive) account is found in such writers as Braithwaite, Carnap, Hempel, Nagel, and Popper; the historicist account (though not under that name) in Feyerabend, Hanson, and Kuhn. I have dubbed the latter view "historicist" because it depends largely upon detailed analyses of historical examples, and tends in various ways toward a conventionalist or historically relativist view of science and its methods. I do not mean to imply, however, that this characterization exhausts the valuable insights to be found in these writers, nor that they hold anything like a common view. For example, Hanson, unlike Feyerabend and Kuhn, was prepared to argue for the existence of a logic of discovery

along lines similar to those I shall pursue here: cf. his "Is There a Logic of Discovery?", *Current Issues in the Philosophy of Science,* ed. H. Feigl and G. Maxwell (New York: Holt, Rinehart & Winston, 1961), p. 20.

2. These remarks occur scattered throughout his writings. Some of the most perceptive are found in an early paper written for the Apostles' Club at Cambridge in 1856: "Analogies. Are There Real Analogies in Nature?", *Life of James Clerk Maxwell,* L. Campbell and W. Garnett (London, 1884), p. 347; and in the first pages of "On Faraday's Lines of Force [1856]" *(FL), Scientific Papers,* ed. W. D. Niven (Cambridge, 1890), 1:155. More extensive methodological discussions, especially of the significance of the generalized Lagrangean formulation of mechanics, are to be found in a series of papers from 1870 to 1879, for example, "Address to the Mathematical and Physical Sections of the British Association" (2:215), "On the Mathematical Classification of Physical Quantities" (2:257), "On the Proof of the Equations of Motion of a Connected System" (2:308), "Action at a Distance" (2:311), "On the Dynamical Evidence of the Molecular Constitution of Bodies" (2:418), "Thomson and Tait's Natural Philosophy" (2:776).

3. G. E. Davie, *The Democratic Intellect* (Edinburgh, 1961), p. 192 ff. I owe this reference to the unpublished Ph.D. thesis of P. M. Heimann, "James Clerk Maxwell, His Sources and Influence" (Leeds, 1970).

4. *Scientific Papers,* 1:155.

5. Maxwell's own discussions of physical analogy have been considered by Joseph Turner in two pioneering papers: "Maxwell's Method of Physical Analogy," *Brit. J. Phil. Sci.* 6 (1955):266; and "Maxwell on the Logic of Dynamical Explanation," *Phil. Sci.* 23 (1956):36. Turner does not recognize, however, the function of analogies in providing inductive arguments in Maxwell or elsewhere. Indeed, he explicitly denies (but without giving reasons) that the method of physical analogy permits inference from one physical system to another which is similar to it in some respects ("Maxwell's Method," p. 238). In this he seems to differ from Maxwell himself, as the remark just quoted from Maxwell in the text indicates.

In another analysis of Maxwell's method of analogy, "Model and Analogy in Victorian Science, Maxwell's Critique of the French Physicists," *Journ. Hist. Ideas* 30 (1969):423, Robert Kargon has, almost alone among recent commentators, noticed that the analogical method is regarded by Maxwell as inductively preferable to the hypothetical and the mathematical. He has, however, interpreted analogy as a second best to the method of deduction from experiments, when the latter is applicable. In this he differs from the interpretation given later in this paper, where an intimate relation is discerned between the analogical and the deductive methods.

6. The most explicit discussion of the last point occurs in *An Ele-*

mentary Treatise on Electricity, ed. W. Garnett (Oxford, 1881), Section 64, and especially Section 113:

The effects of the current of conduction on the electrical state of A and B are of precisely the same kind as those of the current of convection. . . . In the case of the convection of the charge on the pith ball we may observe the actual motion of the ball. . . . But in the case of the current of conduction through a wire we have no reason to suppose that the mode of transference of the charge resembles one of [the] methods [of convection] rather than another. All that we know is that a charge of so much electricity is conveyed from A to B in a certain time.

7. The claim to deduction from experiments occurs not only in Maxwell's more discursive methodological writings, but also throughout his mature electromagnetic theory, for example: "the laws of the distribution of electricity on the surface of conductors have been analytically deduced from experiment" (*FL, Scientific Papers,* 1:155), "the conclusions arrived at in the present paper are independent of this hypothesis of motions and strains in the aether, being deduced from experimental facts." ("A Dynamical Theory of the Electromagnetic Field [1864]" *(DT), Scientific Papers,* ed. W. D. Niven [Cambridge, 1890], 1:564.) "I propose . . . to state [the electromagnetic theory] in what I think the simplest form, deducing it from admitted facts." ("Note on the Electromagnetic Theory of Light [1868]" *(Note),* Ibid., 2:138.) Cf. also Maxwell's correspondence, at the time of the publication of "On Physical Lines of Force [1861/2]" *(PL),* Ibid., vol. 1 and *DT,* in Campbell and Garnett, pp. 246, 255.

8. I. Newton, *Mathematical Principles of Natural Philosophy,* tr. Motte-Cajori (Berkeley and Los Angeles: University of California Press, 1934), p. 398. In this quotation I have omitted the significant and difficult phrase "which admit neither intensification nor remission of degrees," for which, see J. E. McGuire, "The Origin of Newton's Doctrine of Essential Qualities," *Centaurus* 12 (1968):233. This paper and the same author's "Atoms and the 'Analogy of Nature': Newton's Third Rule of Philosophizing," *Studies in Hist. Phil. Sci.* 1 (1970):1, contain detailed analyses of Newton's Rules. For the logic of "deduction from phenomena," see the important series of papers by J. Dorling, "Maxwell's Attempts to Arrive at Non-speculative Foundations for the Kinetic Theory," *Studies in Hist. Phil. Sci.* 1 (1970):229; "Einstein's Introduction of Photons: Argument by Analogy of Reduction from the Phenomena?", *Brit. J. Phil. Sci.* 22 (1971); "Henry Cavendish's Deduction of the Electrostatic Inverse Square Law from the Result of a Single Experiment," presented to a Cambridge Seminar, 1970; and "The Validity of Deduction from the Phenomena," read to the British Society for the Philosophy of Science, 1970.

9. But see Section II below, where "generalization from experience"

is taken to include problematic "experimental identifications" which may give richer content to the terms of Lagrange's equations compared with those of Newton. On this question see an interesting article by T. K. Simpson, "Some Observations on Maxwell's *Treatise on Electricity and Magnetism,*" *Studies in Hist. Phil. Sci.* 1 (1970):249, which appeared too late for me to take explicit account of it in this paper.

10. J. C. Maxwell, *A Treatise on Electricity and Magnetism,* 1st ed. (London, 1873), Section 552.

11. This argument nearly becomes explicit in *Elementary Treatise,* Section 5: "Definition.—*Whatever produces or tends to produce a transfer of Electrification is called an Electromotive Force.*"

12. He describes it in *Treatise on Electricity,* Section 62, where he says it "may be actually true," provided there are no dielectrics with constant less than that of vacuum. He does not explicitly "reject Mossotti's hypothesis," as stated erroneously by Fitzgerald in "M. Poincaré and Maxwell," *Scientific Writings of G. F. Fitzgerald,* ed. J. Larmor (Dublin and London, 1902), p. 284.

13. See for example P. Duhem, *Les Théories électriques de J. C. Maxwell* (Paris, 1902); O. Heaviside, "Electromagnetic Induction and Its Propagation [1885–7]," *Electrical Papers* (London, 1892), 1:434, 477; H. Hertz, *Electric Waves,* 1892, tr. D. E. Jones (London, 1893), Introduction; H. Poincaré, *Electricité et Optique* (Paris, 1901), p. viii and *passim.;* J. J. Thomson, "Report on Electrical Theories," *Report of the 55th Meeting of the British Association* (Birmingham, 1886), p. 125 ff. Among more recent commentators, see A. O'Rahilly, *Electromagnetic Theory* (New York, 1965), first published as *Electromagnetics* (London, 1938), pp. 76 ff.; J. Bromberg, "Maxwell's Displacement Current and his Theory of Light," *Arch. Hist. Ex. Sci.* 4 (1967/8):218; and Maxwell's Electrostatics," *Amer. J. Phys.* 36 (1968):142; P. M. Heimann, "Maxwell and the Modes of Consistent Representation," *Arch. Hist. Ex. Sci.* 6 (1970):171; and "Maxwell, Hertz and the Nature of Electricity," *Isis* 62 (1971):149–157.

14. "[I endeavour] to explain the action between distant bodies without assuming the existence of forces capable of acting directly at *sensible* distances" (1:527, my italics).

15. Cf. *Treatise on Electricity,* Section 61. Since curl $B = (4\pi/c)j + (1/c)\dot{D}$, and any div curl function is identically zero, div $(4\pi/c)j + (1/c)\dot{D}$ is everywhere zero.

16. A. O'Rahilly remarks with approval: "The ordinary working physicist remains convinced that a current really consists of something travelling along the wire. He remains frankly sceptical in face of the paradoxical hypothesis that it is everywhere except in the wire" (*Electromagnetic Theory,* p. 277). J. H. Poynting, in "On the transfer of

energy in the electromagnetic field," *Phil. Trans.* 175 (1884):360, is more sensitively aware of the paradigm switch required to understand the second interpretation:

It is very difficult to keep clearly in mind that this 'displacement' is, as far as we are yet warranted in describing it, merely a something with direction which has some of the properties of an actual displacement in incompressible fluids or solids. . . . It seems to me then that our use of the term is somewhat unfortunate, as suggesting to our minds so much that is unverified or false, while it is so difficult to bear in mind how little it really means.

17. J. H. Poynting, "Transfer of Energy" and "On the Connexion Between Electric Current and the Electric and Magnetic Inductions in the Surrounding Field," *Phil. Trans.* 176 (1885):277; O. Heaviside, "Electromagnetic Induction." Maxwell's first explicit statement of the second interpretation is in *Treatise on Electricity,* Section 111:

The electrification therefore at the bounding surface of a conductor and the surrounding dielectric, which on the old theory was called the electrification of the conductor, must be called in the theory of induction the superficial electrification of the surrounding dielectric. According to this theory, all electrification is the residual effect of the polarization of the dielectric. . . .

In the phenomenon called the electric current the constant passage of electricity through the medium tends to restore the state of polarization as fast as the conductivity of the medium allows it to decay. Thus the external agency which maintains the current is always doing work in restoring the polarization of the medium, which is continually becoming relaxed, and the potential energy of this polarization is continually becoming transformed into heat.

18. For an analysis of the significance of Hertz's experiments in relation to Maxwell's theory, see S. d'Agostino, "Hertz's Discovery of Electromagnetic Waves," forthcoming.

19. J. Bromberg seems to be mistaken in her suggestion that in the second interpretation Maxwell has "cut [his equations] off from the physical ideas [the first interpretation] upon which he had founded them" ("Maxwell's Electrostatics," p. 151), if this implies that the first interpretation has some inductive cogency which the second lacks. At best it may be said that Maxwell's first understanding of the meaning of the displacement current was probably in terms of the first interpretation. But in the absence in most commentators of any distinction between the meaning of different models and their inductive force, it is difficult to know how to interpret judgments such as Bromberg's.

20. This logical point about a multiplicity of analogies gives a rational ground for Maxwell's *pluralism* of interpretations, an aspect of his physics which is remarked upon by d'Agostino, "La Pensée scientifique de Maxwell et le développement de la théorie sur champ électromagnéti-

que dans le mémoire 'On Faraday's lines of force'," *Scientia* 103 (1968):7. It is also enough to put in their right perspective Duhem's many accusations of *contradiction* in Maxwell's work (*Les Théories électriques,* p. 101). A pluralism of models does not imply a self-contradictory theory.

5 Charles Darwin and Nineteenth-Century Philosophies of Science

DAVID L. HULL

University of Wisconsin, Milwaukee

To the extent that it can be distinguished from epistemology, philosophy of science as a separate discipline began in England with the publication of the works of John Herschel (1792–1871), William Whewell (1794–1866) and John Stuart Mill (1806–1873). Herschel and Mill carried on in the empiricist tradition which had preceded them, while Whewell was strongly influenced by the writings of Immanuel Kant.[1] But when these philosophers addressed themselves more directly to questions concerning science, they wrote under the twin spectres of Francis Bacon and Isaac Newton. Everyone knew what science was. Science was Kepler's laws, Newton's theory of universal gravitation, William Herschel's extension of physical astronomy to the sidereal regions, and W. C. Wells's "Essay on Dew" (1818). Everyone was equally sure of the general character of scientific method. It was not the "deductive method" of Aristotle and the scholastics but the "inductive method" expounded by Bacon and practiced by these scientific giants. The task which Herschel, Whewell and Mill set themselves was to provide an explicit and detailed analysis of science consonant with the great achievements in physics which had preceded them.

The classic works in nineteenth-century philosophy of science were all published prior to the appearance of Charles Darwin's *Origin of Species* in 1859—Herschel's *Preliminary Discourse on the Study of Natural History* (1830), Whewell's massive volumes on the

History of the Inductive Sciences (1837) and *The Philosophy of the Inductive Sciences, Founded Upon Their History* (1840), and finally Mill's *System of Logic* (1843). Although Mill learned what little he knew of science from reading Whewell's *History of the Inductive Sciences,* his *System of Logic* was in large part an empiricist attack on Whewell's idealist and rationalist views. Whewell refrained from replying to Mill for six years, then published his *On Induction, with especial reference to Mr. J. Stuart Mill's System of Logic* (1849), just in time for Mill to insert numerous lengthy footnotes in the third edition of his *System of Logic* answering Whewell's objections. Was induction a process of discovering empirical laws in the facts, or does the mind superinduce concepts on the facts? Were the axioms of geometry, especially the parallel line postulate, inductions from experience, whose truth could be decided by scientific investigation, or were they self-evident truths, whose formulation might be initiated by experience but once conceived, were known to be true *a priori?* In this debate, Herschel placed his great influence largely on the side of Mill, going so far as to concur in Mill's analysis of deduction as proceeding from particulars through general propositions to particulars.[2]

One would think that Darwin could not have chosen a better time to develop and eventually publish his theory of evolution. Earlier scientists had to produce their theories without any explicitly formulated body of methodological principles to guide them. These theories in turn were judged by tacit, or at best, half-articulated standards. At last a scientist could develop a scientific theory in full knowledge of the proper methods of science, and when his theory was published, it could be evaluated according to the explicitly stated criteria set out by the most sophisticated minds of his day.

In point of fact, Darwin did read Herschel's *Discourse* in his last year at Cambridge. Along with Humboldt's *Personal Narrative* (1818), it stirred in him "a burning zeal to add even the most humble contribution to the noble structure of Natural Science."[3] While at Cambridge, Darwin knew Whewell for a short time and found him a great converser on grave subjects. Darwin was also impressed by the breadth of knowledge exhibited by Whewell in his *History of the Inductive Sciences,*[4] and eventually prefaced the *Origin of Species* with a quotation from Whewell's Bridgewater Treatise (1833). But Darwin was indignant that someone might find Whewell profound "because he says length of days adapted to duration of sleep in man !!! whole universe so adapted !!! and not men to Planets.—instance of arrogance!!!"[5] Probably Darwin was not overly displeased when Robert Brown, the botanist, responded to praise of Whewell's *History* by sneering, "Yes, I suppose that he has read the prefaces of very many books."[6] Finally, there is no evidence that Darwin read Mill's *System of Logic* prior to publishing his *Origin of Species,* but afterwards he could hardly have helped being influenced indirectly through his colleague T. H. Huxley, who was very much impressed by Mill.[7]

The effect which Darwin's reading in philosophy of science had on him seems to have been mixed. Herschel's *Discourse* was certainly inspirational. Scientists and men of God were marching arm in arm to produce a better world. The reference in the opening paragraph of the *Origin* to "one of our greatest philosophers" was to Herschel. Darwin's care in recording exceptions to his own views[8] stemmed from Herschel's admonitions.[9] In general, however, the confusions and contradictions that pervaded early works in the philosophy of science were merely duplicated in Darwin's utterances on the subject. On the one hand, Darwin looked upon "a strong tendency to generalize as an entire evil"[10] and yet admitted that "I cannot resist forming one on every subject."[11] He distrusted "deductive reasoning in the mixed sciences" and claimed, in the formulation of evolutionary theory, to have "worked on true Baconian principles and without any theory collected facts on a wholesale scale."[12] Yet he objected to the view that "geologists ought only to observe and not to theorize," remarking how "odd it is that anyone should not see that all observation must be for or against some view if it is to be of any service!"[13] Perhaps Darwin had no clearer a conception of science and scientific method than the philosophers of his day, but as we shall see, he was certainly no more confused.

The other side of the coin is even more tarnished. One would think that men who had devoted their lives to an analysis of science would be more adept at evaluating a new and revolutionary theory than the average philosopher or practicing scientist. Whatever justification there might be for this surmise, in this instance it could not have been more mistaken. Whewell rejected the theory out of hand. Herschel and Mill were willing at least to entertain the possibility of species evolving but not by the mechanisms proposed by Darwin.

Darwin was so anxious to hear Herschel's opinion of the *Origin of Species* that he wrote to Charles Lyell asking him to pass on any comments which Herschel might make since "I should excessively like to hear whether I had produced any effect on such a mind."[14] It did not take long for Darwin to discover the effect which he had produced on Herschel. Again to Lyell, he wrote, "I have heard, by a roundabout channel, that Herschel says my book 'is the law of higgledy-piggledy.' What this exactly means I do not know, but it is very contemptuous. If true this is a great blow and discouragement."[15] Herschel expanded on his criticism of Darwin's theory in the 1861 edition of his *Physical Geography of the Globe,* concluding, "We can no more accept the principle of arbitrary and casual variation and natural selection as a sufficient account, per se, of the past and present organic world, than we can receive the Laputan method of composing books (pushed *à outrance*) as a sufficient account of Shakespeare and the Principia."[16] If Darwin would just admit the necessity of "intelligent direction" in evolution and include it in the formulae of his theory, then Herschel "with some demur as to the genesis of man" was "far from disposed to repudiate the views taken of this mysterious subject in Mr. Darwin's book."

Whewell's refusal to grant even this much to Darwin's theory was to be expected. Since he had steadfastly opposed uniformitarian principles in geology, it was unlikely that any argument or assemblage of data would rapidly convert him to the evolution of organic species. Miraculous divine creations, catastrophes and the argument from design were vastly superior to Darwin's speculations. In a letter to a professor of theology, Whewell expressed his opinion of evolutionary theory as follows:

And I may say that the recent discussions which have taken place in geology and zoology do not appear to me to have materially affected the force of the arguments there delivered.[17] It still appears to me that in tracing the history of the world backwards, so far as the palaetiological sciences enable us to do so, all the lines of connexion stop short of a beginning explicable by natural causes; and the absence of any conceivable natural beginning leaves room for, and requires, a supernatural origin. Nor do Mr. Darwin's speculations alter this result. For when he has accumulated a vast array of hypotheses, still there is an inexplicable gap at the beginning of his series. To which is to be added, that most of his hypotheses are quite unproved by fact. We can no more adduce an example of a new species, generated in the way which his hypotheses suppose, than Cuvier could. He is still obliged to allow that the existing species of domestic animals are the same as they were at the time of man's earliest history. And though the advocates of uniformitarian doctrines in geology go on repeating their assertions, and trying to explain all difficulties by the assumption of additional myriads of ages, I find that the best and most temperate geologists still hold the belief that great catastrophes must have taken place; and I do not think that the state of the controversy on that subject is really affected permanently. I still think that what I have written is a just representation of the question between the two doctrines.[18]

And then, of course, there is also the story that no copy of the *Origin of Species* was allowed on the shelves of the library at Trinity College while Whewell was master.

In the face of such rejection by two of the most eminent philosophers of his day, one can imagine Darwin's elation when he discovered that another great philosopher, John Stuart Mill, thought his reasoning in the *Origin* was "in the most exact accordance with the strict principles of logic."[19] Darwin was prepared for the abuse which the content of his theory, especially its implications for man, was to receive from certain quarters, but he was not prepared for the criticism which his methodology was to receive from the more respected philosophers and scientists of his day. Most contemporary commentators tend to dismiss these criticisms as facile, disingenuous and superficial, suspecting that they stemmed more from a distaste for the content of Darwin's theory than from any shortcomings of his methodology, but this dismissal itself is too facile. Certainly repeated invocations of the Baconian method by many of Darwin's critics and even by Darwin himself indicated no great understanding of the actual nature of this method, but nevertheless there were fundamental conflicts between evolutionary theory and the most sophisticated philosophies of science currently popular.[20]

For example, Mill's endorsement was a two-edged sword, and the sharper edge cut deeply into Darwin's own claims for his book. Darwin looked upon the *Origin of Species* as "one long argument from the beginning to the end, and it has convinced not a few able men."[21] He thought that, to some extent at least, he had proved that contemporary species originated from earlier species by evolution through chance variation and natural selection. According to Mill, Darwin had not violated the rules of induction, since the "rules of Induction are concerned with the conditions of Proof. Mr. Darwin has never pretended that his doctrine was proved. He was not bound by the rules of Induction, but those of Hypothesis."[22] And the method of hypothesis was a method of discovery, not justification. Darwin had admirably fulfilled the requirements of one of the methods of discovery, but he had proved nothing! Newton had provided the necessary inductive proof for his theories; Darwin had not.[23] In his last pronouncements on evolution, Mill agreed with Herschel and Whewell that "in the present state of our knowledge, the adaptations in Nature afford a large balance of probability of creation by intelligence."[24]

Professional philosophers of science played a significant role in the reception of evolutionary theory. Unfortunately, this role was largely negative. The purpose of the remainder of this paper will be to discover what it was about the philosophies of Herschel, Whewell and Mill which led them to reject evolutionary theory as set out by Darwin when it was making converts on every side among biologists and laymen alike. Part of the explanation is that Darwin was caught in the middle of a great debate over some of the most fundamental issues in the philosophy of science—the difference between deduction and induction and the role of each in science, the difference between concept formation and the discovery of scientific laws, the relation between discovery and justification, the nature of the axioms of geometry and their relation to physics, the distinction between occult qualities and theoretical entities, and the conflict between purely naturalistic science and the role of God's direct intervention in nature. Before philosophers of science had thoroughly sorted out these issues, they were called upon to evaluate an original and highly controversial scientific theory. But evolutionary theory was not just another scientific theory. It was a theory that struck at the very foundations of the philosophies of science which were being used to judge it. Of the numerous conflicts between evolutionary theory and nineteenth-century philosophies of science, only two will be discussed here—the conflict between teleology and evolution by the mechanisms proposed by Darwin, and the conflict between essentialism and any evolution at all, as long as it was gradual.

TELEOLOGY, CHANCE VARIATION AND NATURAL SELECTION

The two major varieties of teleology popular in Darwin's day were the external teleology of the Christian Neoplatonists and the immanent teleology of Aristotle. The conflict between evolutionary theory and immanent teleology has been examined at length elsewhere and will not be discussed here (Hull 1965, 1967). The doctrine of external teleology entails a universal consciousness ordering everything for the best. As Plato put it, "I heard some one reading, as he said, from a book of Anaxagoras, that mind was the disposer and cause of all, and I was delighted at this notion, which appeared quite admirable, and I said to myself: if mind is the disposer, mind will dispose all for the best, and put each particular in the best place."[25] It was the Christian doctrine of God as the disposer and cause of all that Darwin had to combat.

In his Bridgewater Treatise (1833), Whewell updated the arguments for God's existence which Richard Bentley had set forth in his "A confutation of atheism from the origin and frame of the world" (1693). Like Bentley, Whewell argued that Newtonian physics was not only compatible with the existence of God but also necessitated it. Certain phenomena like the earth traveling in an ellipse around the sun were deducible from Newton's laws. Other phenomena like the sun being luminiferous and the planets opaque, the distances of the various planets from the sun, and the inclination of the earth's axis to the plane of its revolution, were not. As far as Newton's laws were concerned, these latter phenomena were "accidental" features of the universe. Yet they had to be precisely as they were if the earth was to support life. If the earth were further away from the sun, it would be too cold to support life; if it were nearer, too hot. If the earth's axis had not been inclined to its plane of rotation, most of the planet would be either too hot or too cold for habitation. If the earth revolved around the sun more slowly, the seasons would be too long; if more quickly, too short, and so on. Since these phenomena necessary for man's existence did not follow from Newton's laws, either they just happened to coincide or else God was responsible. The chance coincidence of so many beneficial phenomena was extremely implausible. Hence, God was responsible, and if he was responsible for these phenomena, he must exist. The existence of life, especially man, was good, hence, God was good. Chance was eliminated from physical phenomena indirectly by God instituting divine laws and directly by his production of those phenomena which did not follow from these laws.

Like others before him,[26] Whewell entertained a possible alternative explanation for the existence of such a high correlation between the needs of living creatures (especially man) and their environment—the survival of the fit and the perishing of the unfit. Whewell found such explanations inconceivable:

> If the objector were to suppose that plants were originally fitted to years of various lengths, and that such only have survived to the present time, as had a cycle of a length equal to our present year, or one which could be accommodated to it; we should reply, that the assumption is too gratuitous and extravagant to require much consideration.[27]

Mill concurred in Whewell's preference for the argument from design over the principle of the survival of the fittest, even after the publication of *Origin of Species*. "Of this theory when pushed to this extreme point, all that can now be said is that it is not as absurd as it looks, and that the analogies which have been discovered in experience, favourable to its possibility, far exceed what any one would have supposed beforehand."[28]

The facility with which Herschel, Whewell and Mill could demand the exact verification of scientific hypotheses and the exclusion of occult qualities from science on the one hand while on the other asserting God's direct intervention in natural phenomena is nothing less than schizophrenic. All three men had argued for the elimination of the miraculous intervention of God in the material world and the expansion of scientific law to cover all physical phenomena. As Mill expressed himself on the issue, there were two concepts of theism, one consistent with science, one inconsistent. "The one which is inconsistent is the conception of a God governing the world by acts of variable will. The one which is consistent, is the concept of a God governing the world by invariable laws."[29] The more that empirical phenomena could be shown to be governed by secondary causes acting in accordance with divinely instituted laws, the more powerful and omniscient God was shown to be. In the quotation which Darwin had selected to introduce the *Origin of Species,* Whewell urges, "But with regard to the material world, we can at least go so far as this— we can perceive that events are brought about not by insulated interposition of Divine power, exerted in each particular case, but by the establishment of general laws."[30] What these pious men did not perceive was that by removing God as an active agent and relegating him to the position of the divine author of immutable laws, they were preparing the way for his total expulsion from science. Like Kant's *Ding an sich,* he was becoming remote, obscure, unknowable, somehow underlying everything and very important, but of no conceivable consequence for any particular scientific investigation.

All of these authors also recognized limits to natural laws, limits which were conveniently just on the other side of currently accepted laws. For example, Whewell argued that theology had to be assiduously excluded from the dynamic sciences but was a necessary adjunct to the historical sciences. "The mystery of creation is not within the legitimate territory of science."[31] Herschel agreed, "To ascend to the origin of things, and speculate on the creation, is not the business of the natural philosopher."[32]

Nineteenth-century scientists are often belabored for believing in special creation, the miraculous flashing together of elemental atoms to produce fully formed organisms or their eggs, but the other alternative, spontaneous generation, seemed even less plausible. "Spontaneous generation" implied much the same process only through some unknown though natural agency. The introduction of evolutionary theory did not change the situation much. If evolution of some sort is admitted, the problem is reduced in scope and pushed back into the distant past, but it is not eliminated. Either the first members of the original species were specially created or else they were spontaneously generated. By the middle of the nineteenth century the advocates of spontaneous generation had retreated from the spontaneous generation of mice and such to the spontaneous generation of microbes and green matter, a position similar to that held by Lamarck in the preceding century. In Darwin's day the spontaneous generation of an elephant or a mouse was thought to be absurd. The spontaneous generation of a microbe or an egg by some unknown laws was not quite so absurd. The major advantage which Darwin's theory had over its rivals in this matter was that it required the spontaneous generation of one or a few very simple beings a long time ago. Coincidentally, just when Pasteur was proving that contemporary microbes were not generated spontaneously, evolutionary theory required the spontaneous generation of ancient microbes.

After the appearance of the *Origin,* Darwin was thoroughly denounced for implying that the flashing together of elemental atoms had been the accepted doctrine among biologists on the origin of species. Biologists such as Richard Owen claimed to have presumed, though not speculated on, a naturalistic explanation for the development of species, a mechanism which in Owen's case was embarrassingly similar to that outlined in the *Vestiges of Creation.* Earlier biologists such as Linnaeus had not been outright special creationists. Rather they had suggested some special creation and some natural development. Certain species were specially created. Others developed from these original species by hybridism, paedogenesis, or some such mechanism. These bastardizations were somewhat more popular than Tycho Brahe's model for the universe, part geocentric, part heliocentric, but not much. Regardless of what they might have thought on the subject, one thing is certain—most serious British scientists in Darwin's day studiously avoided the question of the origin of species. They were encouraged in this conspiracy of silence by the philosophies propounded by Herschel, Whewell and Mill. Certain questions were beyond the reach of science.

It is often said that evolutionary theory brought an end to the inclusion of God as a causal factor in scientific explanations. A more accurate characterization is that Darwin's theory demonstrated conclusively that this day had already passed. The architects of this demise were not atheistic materialists but pious Christians who thought that they were doing religion good service by interposing invariable laws between God and the empirical world. Everything except the origin of the universe and organic species was governed by natural laws. It was only a matter of time until the boundaries of science were enlarged to include these phenomena as well. Even Herschel and Mill were willing to abandon the limitations which they had previously placed on the reign of natural law. They were willing to entertain naturalistic explanations for the introduction of living species into the world, but not Darwin's explanation! Whewell refused to go even this far. As we will see, Whewell's refusal to accept the gradual evolution of species regardless of the mechanism may well have stemmed from his understanding the issues at stake more thoroughly than Herschel and Mill.

Herschel and Mill found Darwin's mechanism for the evolution of species objectionable on two counts. First, chance variation sounded too much like the absence of law. The introduction of new species might be by law but not the law of higgledy-piggledy. Darwin repeatedly stated that by "chance variation" he did not mean to imply the absence of laws governing these variations. Rather he thought that whatever these laws might be, they were unknown and to the best of his knowledge were not teleological in nature.[33] There appeared to be no correlation between the variations which a species might need and those it might get. When most organisms that are born die without leaving issue and the vast majority of species that evolve become extinct without evolving into new species, it was difficult to argue for much in the way of divine guidance in these matters. As Darwin objected to Lyell,[34] "If you say that God ordained that at some time and place a dozen slight variations should arise, and that one of them alone should be preserved in the struggle for life and the other eleven should perish in the first few generations, then the saying seems to be mere verbiage. It comes to merely saying that everything that is, is ordained."

The second reason that Darwin's theory was objectionable to Christian teleologists was the God that it implied. If God created the universe, then one should be able to infer his character from the order evident in his creation. If the universe is a perfectly running machine with a place for everything and everything in its place, then one type of mind is implied. This was the vision of the universe that motivated

Herschel, Whewell and Mill to expand the domain of scientific law. God could have constructed the world so that species evolved by chance variation and natural selection, but that kind of God did not seem especially worthy of love and veneration. The God implied by evolutionary theory and a realistic appraisal of the organic world was capricious, cruel, arbitrary, wasteful, careless and totally unconcerned with the welfare of his creations.

Friends and enemies alike urged Darwin to include a little divine guidance in the laws of evolutionary theory. Darwin replied that there was no more need to mention intelligent direction in the formulae of his theory than there was to include it in Newton's theory. "No astronomer, in showing how the movements of planets are due to gravity, thinks it necessary to say that the law of gravity was designed that the planets should pursue the course which they pursue. I cannot believe that there is a bit more interference by the Creator in the construction of each species than in the course of planets."[35] Reference to God had become as otiose in biology as it had long been in physics.

EVOLUTION, NATURAL KINDS AND SCIENTIFIC LAWS

Alvar Ellegård (1958) in his outstanding book on the reception of evolutionary theory in nineteenth-century England views the controversy over the evolution of species as a conflict between empiricist and idealist philosophies of science. The empiricists promoted the acceptance of evolutionary theory. The idealists were the obscurantists. Our story would be much neater if this were true. In England, at least, it is not. Empiricists and idealists alike were opposed to species evolving and with good reason. Evolution by chance variation and natural selection conflicted with teleology, but with sufficient modification these philosophies could do without teleology. Any evolution at all, regardless of the mechanisms, if it were gradual, conflicted with the essentialist notion of natural kinds, and none of these philosophies could do without natural kinds. The empiricists and idealists attributed a different ontological status to natural kinds, but both agreed that their existence was absolutely necessary if knowledge was to be possible. Thus Peirce (1877) was correct when he observed, "The Darwinian controversy is, in large part, a question of logic," and Dewey (1910) when he concluded, "The real significance of Darwinian evolution was the introduction of a new 'mode of thinking,' and thus to transform the 'logic of knowledge.' "

One would expect an idealist like Whewell to presuppose the existence of natural kinds. In fact, as far as species of plants and animals are concerned, he did not even bother to argue the issue. He merely referred to the opinion of the most eminent physiologists of his day that indefinite divergence from original types was impossible. For Whewell,[36] *"Species have a real existence in nature,* and a transition from one to another does not exist." The impossibility of evolution was not a generalization from experience but a necessary prerequisite for knowledge. "Our persuasion that there must needs be characteristic marks by which things can be defined in words, is founded on the assumption of the *necessary possibility* of reasoning."[37] If species evolved gradually, then no one set of necessary characteristics would exist which was sufficient to divide an evolving lineage into discrete species. On Whewell's philosophy, if Darwin's theory were true, knowledge would be impossible—not a small drawback. (Hull 1965, 1967)

Contrary to Ellegård's belief, the empiricist philosophies of Herschel and Mill equally depended on the existence of discrete natural kinds. Mill, for example, distinguished between two types of uniformities in nature—those of succession in time and those of coexistence. The former were expressed in causal laws; the latter in definitions and classification. Mill's four (sometimes five) methods of induction were designed to discover successions of kinds of events in time, on the assumption that these kinds of events had already been discovered. But as Whewell was happy to point out, Mill's methods "take for granted the very thing which is most difficult to discover, the reduction of phenomena to formulae."[38] In the last analysis, Mill justified causal laws by reference to the law of universal causation. He had no comparable justification for his uniformities of coexistence. Even so, he still maintained that natural kinds were "distinguishable by unknown multitudes of properties and not solely by a few determinant ones—which are partitioned off from one another by an unfathomable chasm. . . ."[39] The universe, so far as known to us, is so constituted, that whatever is true in any one case, is true in all cases of a certain description; the only difficulty is, to find the description.[40] "Kinds are classes between which there is an impassible barrier. . . ."[41]

Although the empiricist ontologies of Herschel and Mill did not require the existence of discrete natural kinds, their logic of justification did. Proof was to be supplied by eliminative induction. "Either A, B or C can cause E. A and B are absent. Hence, C must cause E." In order for this inference to afford absolute certainty, all of the alternative causes for a particular kind of event must be specified and all but one eliminated. It is interesting to note that the one kind of induction which Darwin grew to distrust was in fact exclusive induction. Early in his career he tried to explain the parallel shelves (or roads) that rimmed the sides of Glen Roy (1839). He reasoned that the parallel roads were the former shores of lakes or arms of the sea. If they were the shores of a series of lakes, then huge barriers had to be erected and removed successively at the mouth of the glen. Since Darwin could not see how such huge barriers could be moved about, he eliminated the lake hypothesis. Hence, the shelves were former shores of the sea produced as the land gradually rose above sea-level. Shortly thereafter Agassiz's glacier theory provided the barriers required by the lake hypothesis. In writing to Lyell, Darwin exclaimed, "I am smashed to atoms about Glen Roy. My paper was one long gigantic blunder from beginning to end. Eheu! Eheu!"[42] Later he observed that "my error has been a good lesson to me never to trust in science to the principle of exclusion."[43]

Here was Darwin refusing to trust the principle of exclusion—the very mode of inference that empiricist philosophers were touting as the only method of proof in the empirical sciences! But perhaps the fault lay not with the principle but with Darwin's application of it. In this instance there might be some justification for the claim, but when we turn to species of living creatures, evolution necessarily precludes successful application of induction by complete elimination. Eliminative induction requires the existence of a finite number of sharply distinguishable natural kinds. But if Darwin's theory were true, then eliminative induction could never be applied to living species as parts of temporal continua, since they were neither discrete nor denumerable.

The conflict was more serious than it might appear. The chief examples of natural kinds in natural science had always been geometric figures, species of plants and animals and, running a poor third, physical elements. If organic species did not form natural kinds, then doubt was raised about other natural kinds. Perhaps parallel lines could meet. One might even be able to transmute lead into gold. Scientific laws themselves might evolve. If Darwin's theory were true, then living species did not form natural kinds. If living species did not form natural kinds, then "All swans are white," even if true, could not be a scientific law. No wonder empiricists like Huxley were disposed toward saltative evolution.

The preceding inferences, however, do not follow as automatically as nineteenth-century empiricists thought. They believed that scientific laws were necessarily universal in form. Approximations were just stages on the road to something better—true universal generalizations.[44] Anything less implied an indeterministic universe and made experimental verification impossible. Mill admitted that in many instances recourse to approximations was necessary, but only because the phenomena had not been correctly reduced to natural kinds. Once natural kinds had been discerned, universal correlation was guaranteed.[45] If species of plants and animals did not form natural kinds, then biological laws about them might forever remain approximations. Peirce[46] did not find this conclusion so abhorrent. Just as Maxwell had applied the statistical method and the doctrine of probabilities to the theory of gases, "Darwin, while unable to say what the operation of variation and natural selection in every individual case will be, demonstrates that in the long run they will adapt animals to their circumstances." The question posed by Darwin's theory to nineteenth-century philosophies of science was whether statistical laws and classes defined by statistically covarying properties could function in science, not as temporary stopgaps, but as permanent features of science. (For further discussion, see Hull, *Philosophy of Biological Science,* 1972.)

Herschel and Mill did not agree with Whewell on very many issues, but it was those issues on which they did agree that made their acceptance of evolutionary theory impossible. If evolutionary theory was to be accepted, certain basic changes had to be made. From Lyell, Asa Gray, Herschel, Mill and even Peirce to Teilhard de Chardin, philosophers and biologists have urged a little bit of direction in evolution. From Huxley to de Vries, Goldschmidt, Schindewolf, and Goudge they have hoped that evolution might prove to be saltative. The motivation for the first objection was to salvage some remnant of teleology; for the latter to retain the essentialist mode of definition and all that it entailed. Darwin yielded to neither temptation.

NOTES

1. For an extensive discussion of the metaphysics and epistemologies of such philosophers as Descartes, Locke, Hume, Berkeley, Leibniz, and Kant and their relation to science, see Gerd Buchdahl's *Metaphysics and the Philosophy of Science* (Oxford: Basil Blackwell, 1969).

2. See Herschel's review of Whewell's *History and Philosophy of Science* (1841) and his comments on Mill in his review of Quetelet's essays (1850), both in John Herschel, *Essays from the Edinburgh and Quarterly Review with Addresses and Other Pieces* (London: Longman's, Green, 1857).

3. Charles Darwin, *The Autobiography of Charles Darwin, 1809–1882,* ed. Nora Barlow (London: Collins, 1958), p. 67.

4. Ibid., p. 66.

5. Charles Darwin, "Darwin's Notebooks on Transmutation of Species," ed. Gavin De Beer, *Bulletin of the British Museum (Natural History) Historical Series* 2:23–200, Third Notebook, p. 134.

6. Darwin, *Autobiography,* p. 104.

7. See Huxley's *Lectures and Essays,* note, and his *Lay Sermons,* both in T. H. Huxley, *Collected Essays* (London: Macmillan, 1893–94; reprinted, New York: Georg Olms Verlag Hildesheim, 1970), vols. 1, 9, pp. 54 and 95 respectively.

8. Darwin, *Autobiography,* p. 123.

9. John Herschel, *Preliminary Discourse on the Study of Natural Philosophy* (London: Longman, Rees, Orme, Brown & Green, 1830). A facsimile of the first edition (New York: Johnson Reprint Corporation, 1966), p. 165.

10. Charles Darwin, *More Letters of Charles Darwin,* ed. Francis Darwin and A. C. Seward (London: Murray, 1903), 1:39.

11. Darwin, *Autobiography*, p. 141.

12. Ibid., p. 119.

13. Darwin, *More Letters*, 1:195.

14. Charles Darwin, *The Life and Letters of Charles Darwin, including an autobiographical chapter*, ed. Francis Darwin, 3 vols. (London: Murray, 1887), 2:26.

15. Darwin, *Life and Letters*, 2:37.

16. Von Baer expands on this reference to Swift in his *Zum Streit über den Darwinismus* (1873), *Augsburger Allgemeine Zeitung*, pp. 1986–1988, translated in *Darwin and His Critics*, ed. D. L. Hull (Cambridge: Harvard University Press, 1972).

17. The reference is to his own *Indications of the Creator* (1847), a series of extracts from his *History* and his *Philosophy of the Inductive Sciences*.

18. Isaac Todhunter, *William Whewell, D.D.* (London: Macmillan, 1876), 2:433–4.

19. Darwin, *More Letters*, 1:189–190.

20. For further discussion of these issues, see Hull, *Darwin and His Critics*.

21. Darwin, *Autobiography*, p. 140.

22. John Stuart Mill, *A System of Logic, Ratiocinative and Inductive, Being a Connected View of the Principles of Evidence, and the Methods of Scientific Investigation* (1843), new impression, 8th ed. (London: Longmans, 1961), p. 328.

23. Ibid., p. 323 and H. S. R. Elliot, ed., *The Letters of John Stuart Mill* (London: Longman, Green and Company, 1910), 2:181.

24. John Stuart Mill, *Three Essays on Religion* (1874) (New York: The Liberal Arts Press, 1958), p. 172.

25. Plato, *Phaedo*, trans. Benjamin Jowett (Oxford: The Clarendon Press, 1871), st. 96–99.

26. Aristotle, *Physics*, ed. David Ross (Oxford: Oxford University Press, 1937), bk. II, ch. 8.

27. William Whewell, *Astronomy and General Physics considered with reference to natural theology* (1833) (Reprint, Philadelphia: Carey, Leo & Blanchard, 1936), vol. III of the Bridgewater Treatise, p. 27.

28. Mill, *Three Essays*, p. 172; see also, *Letters of John Stuart Mill*, 1:236.

29. Mill, *Three Essays*, p. 135.

30. Whewell, *Astronomy*, p. 182.

31. William Whewell, *The Philosophy of the Inductive Sciences, founded upon their history* (1840) 2nd ed. (London: J. W. Parker, 1847), 3:309.

32. Herschel, *Discourse*, p. 38.

33. Charles Darwin, *On the Origin of Species by Means of Natural Selection, or the Preservation of Favoured Races in the Struggle for for Life* (1859). A facsimile of the first edition (Cambridge, Mass.: Harvard University Press, 1966), pp. 74, 170, 364.

34. Darwin, *More Letters,* 1:172.

35. Darwin, *More Letters,* 1:154.

36. Whewell, *Philosophy,* 3:626.

37. Ibid., 1:476.

38. Ibid., 1:263.

39. Mill, *Logic,* p. 80.

40. Ibid., p. 201.

41. Ibid., p. 471; see also p. 379.

42. Darwin, *More Letters,* 2:188.

43. Darwin, *Autobiography,* p. 84.

44. Mill, *Logic,* p. 387.

45. Ibid., p. 388; and John Stuart Mill, *An Examination of Sir William Hamilton's Philosophy, and of the Principal Philosophies and Questions discussed in his writings* (1865) (Boston: W. V. Spencer, 1968), 2:308.

46. C. S. Peirce, "The Fixation of Belief," *Popular Science Monthly* 12:1–15 (1877):94.

6 The Genesis and Structure of Claude Bernard's Experimental Method

JOSEPH SCHILLER
Cercle d'Etude Historique des Sciences de la Vie

At three different times, history decided that science should meet the experimental method. Physics was first at the meeting, with Galileo acting as mediator, followed by chemistry, with Lavoisier as the intercessor. It was not until two and a half centuries after the first encounter, and eighty years after the second, that the third encounter between life sciences and the experimental method took place. Nor could it have taken place before the middle of the nineteenth century, when the proper conditions came into existence. The conditions may be summarized as the progress accomplished by physics and chemistry, which provided tools applicable to the study of life processes, the great improvements in the practice of vivisection, and the substitution of a new concept of function for the old *animata anatome* (the use of the parts). The time was ripe for amalgamating the conditions into a new synthesis; its outcome was the experimental method. Had it not been for the existence of those conditions, of which Bernard took the fullest advantage, his name might have fallen into oblivion. This is the meaning of his saying "I am a product of my time."[1] It is also known with the greatest precision where the rendezvous between Claude Bernard and the experimental method took place. It was in the laboratory, and the precision of this location has a definite meaning. It indicates the only objective way to be followed in the study of the genesis and structure of the experimental method consistent with its historical development. This approach raises the problem of the relationship between Claude Bernard's method and philosophy.

133

There are nowadays two main approaches to historiography, intellectual history and factual history. With its emphasis on concepts and neglect of facts, intellectual history is not a suitable approach for the understanding of this page of history written by Bernard.

No scientist is a greater victim of the use of concepts than Claude Bernard. One author (the term historian will not be used in this connection) claims that he was able to demonstrate by their use that Bernard "was always unfaithful to his own principles and his own method,"[2] a statement made in connection with the most far-reaching discovery in nineteenth-century physiology and biochemistry, hepatic glucogenesis, a fundamental work in the establishment of the experimental method in Bernard's own confession. It is by virtue of the concept of the wholeness of the organism that Bernard's notion of disease based on irrefutable clinical and experimental findings is disputed[3] while on account of the same concept the antivivisectionist Cuvier is considered as having influenced the course taken by physiology in the nineteenth century,[4] a claim that defies the ABCs of physiology and of its history.

It is in the perspective of the history of ideas that unwarranted comparisons are made between the French and the German physiologists of the nineteenth century, opposing the "materialistic mechanism" of the latter to the "vitalistic materialism" of the former, in complete disregard of the nature of the problems under investigation, which belong to non-comparable categories, and the means used for their solution.[5] If we ignore the practitioners of vivisection, such as du Bois-Reymond and Helmholtz, German physiologists selected their problems according to the possibility of applying mathematical calculations and physical measurements; the results were foreseeable from the already known laws of physics. French physiologists worked on the living animal and accepted the possibility that they might be caught by the novelty of an unexpected phenomenon subject to biological laws that remained to be elucidated. Under those circumstances the experimental method became a necessity for the life sciences and could evolve. This is the meaning of Bernard's saying that he discovered phenomena he was not searching for, while Helmholtz knew what to expect in his investigations.

The extension of the conceptual opposition between French and

German physiologists resulted in the confusion of physics with chemistry as methodological tools, of anatomy with vivisection; physics and vivisection were put on an equal footing as means of investigation.[6]

The above criticism is not intended to deny the importance of concepts, but to remind us that they acquire value only in so far as they can be related to definite facts, that is, can be brought down to earth. Bernard first made his discoveries and acquired data, and only years later did he meditate on the details, both manual and intellectual, to find out how he operated. Concepts like "internal environment" or "internal secretions" were the outcome of previously discovered facts.

There is no questioning of the right of philosophers to share Bernard's heritage. The experimental method is part of the patrimony both of philosophers and of physiologists. As Henri Bergson rightly stated: "What philosophy owes to Claude Bernard is, first of all, the theory of the experimental method."[7] It should be underlined, however, that Bernard himself was not concerned with a theory of the method but with the method as applicable to physiology. This limitation to Bernard's contribution to philosophy did not gain general acceptance. Because of his intellectual stature, Bernard became the winner's prize in fights between the various philosophical schools as each one tried to capture him for its own benefit—spiritualists, neo-vitalists, agnostics, mechanists, positivists, and atheists—at a time when the impact of science on human activities was becoming predominant. Zola's "experimental novel" was but one manifestation of the trend. A typical example is the medical philosopher Chauffard, who, though he attacked Bernard in his life time for preaching the experimental method and determinism,[8] adopted him after his death, making of him a spiritualist who believed in the existence of a "free intelligence," and meditated on the problem that torments the human mind, that is, the truth that escapes the interrogation of physiology.[9] Another philosopher, Chevalier, who edited Bernard's unpublished *Philosophie,*[10] stated that Bernard, in addition to being the codifier of the experimental method, was also a "metaphysician." "This metaphysic," he said, "is not superimposed on his science; it inspires it."

Such claims could be easily discarded as the products of imaginative minds were it not for the support they find in Bernard's own
writings. They are important to consider because they are related to
scientific facts to which Bernard's own experimental method is inap-
plicable. These facts concern the "directive idea" that in Bernard's
opinion determines well in advance, according to a pre-established
plan, the specific form that an animal will take. A "legislative force"
directs the vital evolution and explains the order and harmony of the
ontogenic development of the organism. The "legislative force" is a
metaphysical force, different from the materialistic "executive force"
which falls within the realm of the experimenter because of its
physico–chemical nature. Historians (Delhoume, Olmsted, Temkin)
have recorded it as a contradiction in Bernard's scientific attitude,
and so have scientists who were annoyed if not vexed by it (Dastre,
Gley, J.-P. Faure, H. Roger) and who tried to save Bernard from
the accusation of neo-vitalism. At first sight it is difficult to reconcile
Bernard's "directive idea" with certain statements on method—for
example, "the experimental method is that of the free thinker,"[11]
"the experimental method will establish the materiality of phenomena"[12]—and with his hope that life could be artificially created in the
laboratory. Here was a basic phenomenon that escaped the experimental method and evoked the first cause, whose futility Bernard
never missed decrying; even the highly praised immediate causes
were inaccessible to investigation. The strangest part of the story is
that no one has paid attention to the context of the problem. Bernard's chosen example was as concrete as any materialistic scientist
might wish it to be, the developing egg.[13] Bernard was confronted
with the problems of heredity and morphogenesis at a time when
Mendel's laws were still unknown and experimental embryology un-
born. He stated that the legislative force is the "atavistic vital force."
The explanation for his use of the term "atavism" instead of the
term "heredity" may be found in Littré's celebrated dictionary, which
defines two meanings of the term "atavism": physiological, resemblance to ancestors; botanical, tendency of hybrids to revert to the
primitive type. As for the term "heredity," it had mostly, though not
exclusively, a juridical connotation. Claude Bernard misstated too

early a problem the scientific validity of which was proved in the twentieth century.

The way is now clear for a discussion of the experimental method. The paternity of the experimental method has been denied to Claude Bernard on conceptual grounds. The expression was used before him,[14] and since it is axiomatically accepted that the term includes the concept, the honor belongs to his predecessors. Bernard did not pretend to this paternity any more than Darwin pretended to that of evolution. Of the three constituents of the experimental method, observation, hypothesis, and experimentation, scattered fragments may be found in the medical and scientific literature of the seventeenth and eighteenth centuries (Gassendi, Perrault, Chaillou, Thierry, Quesnay, Diderot, Zimmerman, Sénebier). There was constant confusion, however, between experience, gained by learning from everyday life, and experiment, meaning a test explicitly devised to prove or disprove a fact or an idea. The term "experiment" was interchangeable with "observation," and reasoning got the lion's share by supplementing both with working hypotheses and conclusions that reflected animism, vitalism, or iatromechanism. Technical procedures were unavailable; chemistry and physics were not sufficiently advanced to provide tools usable on the living. Even vivisection, though widely employed, was not used to the best of its possibilities because of its limited aim, that of legalizing functions. Qualitatively and quantitatively, Bernard's method differed from those of his predecessors. Any influences exerted on Claude Bernard should be searched for elsewhere.

By the time Bernard appeared on the stage of scientific research, physiology had known what he called a *Renaissance*[15] born under the impact of three scientists who obviously influenced him: Lavoisier, Bichat, and Magendie, to whom Descartes should be added. They created the scientific climate in which Bernard elaborated the experimental method. A perusal of their contributions sheds light on it.

Lavoisier demonstrated the unity of chemical laws independently of the state of matter, organic or inorganic, as revealed by the identity of the processes of respiration and combustion. It results that there are not two chemistries, one of the living and one of the inanimate; the same analytical quantitative procedures are applicable to both kingdoms. The theory and practice of physiology found a solid ground on which Bernard could build the experimental method.

Another notion that Bernard owed to Lavoisier, which received little attention, is what is called "metabolism" in modern terminology, assimilation and disassimilation, the two facets of the phenomenon of nutrition, considered since Aristotle as the very characteristic of life, both animal and vegetable. Lavoisier's idea contrasted with that of Bichat, who made of those two phrases two antagonistic phenomena that oppose vital to physical forces. Bichat's position was generally accepted by his contemporaries. Lavoisier's played a fundamental role in Bernard's conception of general physiology; it materialized in his discovery of hepatic glucogenesis[16] with concomitant production and destruction of glycogen.

The identity of chemical laws also reflected the influence of Descartes. Bernard considered him a true scientist, "an active philosopher," like Leibniz but unlike Kant and Schelling. It was not the imaginary physiology elaborated in *De l'homme* that impressed Bernard, but the idea that the same mechanical laws apply to man-made machines and to the human machine. The unity of physical laws parallels that of chemical laws. Another contribution by Descartes is the independence of the machine from any metaphysical force. Descartes deprived animals of souls, directors of the organisms. Animals are automats, but automatism implies the constancy of response under constant conditions, the *sine qua non* for experimental investigation. Laboratory work does not escape the impact of Descartes's machine, since any laboratory experiment is devised on the analogy of a machine that is successively disturbed and repaired. "Descartes's conception dominates modern physiology. Living beings are mechanisms," said Bernard.[17] But the most significant contribution in the elaboration of the experimental method was Descartes's "scientific doubt," which translated into laboratory language means the controlled experiment, the cornerstone of Bernard's method. On methodological grounds Brunetière and Henri Bergson establish a link between Bernard's *Introduction* and Descartes's *Discours de la méthode*.

Bichat was a physiologist and an anatomist, and paradoxically it was the latter who influenced Bernard, not the former. His experiments were coarse, not analytical, limited in scope, and aimed at establishing a hierarchy of organs on the wrong assumption that they direct the activities of the whole organism. On the contrary, his science of tissues (general anatomy, renamed histology in 1819) reduced the complexity of organs to simpler structures, a tendency that would find its completion with the discovery of the cell. Similar tissues formed dissimilar organs. On the dynamic side, he reduced functions to properties and ceased to treat them as the privilege of definite organs, making of them a common mark of all parts of the organism. What Bichat's physiology centralized in organs, his general anatomy decentralized in tissues. It is this decentralization that Bernard retained because it opened the possibility of studying the minute parts of the organism in their basic manifestations. His work with curare reflects this approach.

The strongest and most direct influence was Magendie, who determined the turn taken by Bernard's life. If Lavoisier and Bichat influenced his thoughts, Magendie influenced his way of thinking and his way of acting by proclaiming the inseparability of mind and hand. Magendie's greatest merit was to recognize that the only way to the study of the living is experimentation, which had proved to be so successful in the physical sciences, and to make physiology its tributary. Bernard called his influence "tremendous and salutary."[18] He was probably the one "to conjugate the verb to experiment" more than anyone else; his mind was analytical, directed against systems; his hand was expert. He crusaded for the use of vivisection, because the problem of the living can be solved only in the living, and of physical and chemical procedures in the exploration of animals, including man. He worked at a time when he was alone in an antagonistic world, satisfied to deduct function from dissected structure and life from death, while vitalism provided for the underlying mechanism. Above all, Magendie substituted demonstration for reasoning, promoted the experimental method in physiology,[19] and insisted on the fundamental notion of determinism,[20] the only way to certitude.

After reading such eulogy written in Bernard's own hand, one may ask what was left for Bernard himself in the elaboration of the experimental method. The answer is simple—the fundamentals. Bernard took from Lavoisier and Bichat, and then left them behind. In the case of Magendie, he transcended his master, suturing the parts and filling the gaps, and made of an experimental method an infallible working instrument. With respect to any influence to which Bernard was subjected, one general remark is fundamental. It had to pass the test of adequacy by not conflicting with the experimental method. That remained the proof of the pudding.[21]

The *Renaissance* of physiology, like its historical model, had its negative aspects beside its positive accomplishments. Confusion, intellectual obscurity, cloudiness, and misuse of technical procedures ruled out the possibility of the construction of a valid experimental method. It is important to stress that Bernard's awareness of the situation was the result of obstacles that he and his contemporaries met in their laboratory work. The obstacles were ideological, but also concerned research habits and laboratory procedures. A few examples will indicate the problem.

Bichat claimed that manifestations in the living are unstable, that they change a hundred times a day in every organ, that they are unpredictable;[22] and he contrasted them to the regularity and stability of the inorganic world. The search for the underlying mechanism became illusory: "The how of things matters little, the fact alone should suffice."[23] The explanation raised the intervention of the inevitable and capricious first cause, and Bichat's refuge was vitalism. The fugacity of living phenomena, the impossibility of seizing them before they vanish, had seized its victims among the physiologists of the first half of the nineteenth century. Johannes Müller, Escherich, and Tiedemann abandoned physiology for this reason and shifted to anatomy, the dissection of the dead offering a certitude unknown in experimentation on the living. Those physiologists, like Bichat, could not overcome the obstacle represented by the supposed spontaneity of living matter, a dominant problem in physiology, that originated with Francis Glisson (1674) and became acute with Haller (1752). Bernard's reaction was that he would not be "discouraged by the present uncertainties of science."[24] A phenomenon is sudden in its manifestation, not in its elaboration, which may be unveiled by applying a correct method of investigation. The problem was to find a way of fixing the phenomena before they vanished, to rouse, arrest, and repeat them at the will of the experimenter, to convert uncertainties into certitudes. It was to be the triumph of the experimental method.

The denial of obvious findings by opposing other findings to them was a general attitude in Bernard's time. Hence, endless polemics ensued. Bernard used to give an example that greatly impressed him, the controversy between Magendie and Brodie concerning the action of bile on the digestion of fats and the formation of chyme. Both had tied the bile duct. Magendie had found no effect whatever, while Brodie had found the opposite result. Who was right? Both were right. Magendie worked on the dog and Brodie on the cat. In the dog, bile and pancreatic ducts are separated; in the cat, they run closely together. Brodie had unwittingly tied both ducts, thus depriving the animal of the pancreatic secretion in addition to bile. The conclusion reached by Bernard from this and similar experiments is fundamental in the elaboration of the experimental method. A proven fact cannot be negated by another proven fact that opposes it, because nature does not contradict itself. The difference in response should be sought in the respective experimental conditions.

A corollary is the idea that a number of negative results do not disprove one positive result; Bernard spoke authoritatively from his own experience. The puncture of a definite spot in the medulla oblongata resulted at the first attempt in the production of what he called improperly "experimental diabetes." Eight successive attempts failed, until he definitely located the spot. Had he started with the failures he would not have discovered the phenomenon.[25] Here again, the experimental conditions had to be minutely established in order to be reproducible, and the notion of experimental conditions becomes Bernard's beneficent obsession.

In order to adopt such an uncompromising attitude in experimentation, it is necessary to be deeply convinced of the inherent stability of the properties of the organism in spite of their apparent instability. The following example proves that this attitude is a prerequisite for their disclosure. The physiologist Vulpian found that the injection of toad venom in the frog causes death by arrest of the heart. It is innocuous to the toad itself, however. This conclusion was rejected by Bernard, who showed that by increasing the dose of venom he could also kill the toad by the same mechanism. The experiment appeared very simple at a time when the possibility of immunological reactions did not burden the mind of the investigator. The important point is to see what determined Bernard to increase the dose. If the venom arrests the heart in the frog and not in the toad, it would mean that the same structure, the heart fiber, reacts differently to the same cause, implying a difference in properties from one animal species to another. This is inconceivable; it defies science. The same response should be obtained from the same structure under identical conditions. Otherwise the existence of laws, the very foundation of science, must be denied. The difference resides in the experimental condition, not in the properties of the organ that represent the elements of stability. The problem is quantitative, not qualitative, and the simplicity of the experimental conduct contrasts with the subtlety of the thought.

Technical obstacles were another source of controversy because of the discrepancies in the results obtained by various investigators. Claude Bernard knew his own misadventures. For instance, his sugar determinations in the blood varied with the procedure used, biological (fermentation), physical (rotatory power), chemical (alkaline copper reagent). They lacked consistency. At first, figures were too high; later he repeated all previous determinations, his and those of other investigators, and corrected them by applying what he called "experimental criticism," that is, by submitting them to all possible verifications. It may seem incredible today, when amounts of 0.1–1 cc. of blood are used, that in Bernard's time amounts of 25 to several hundred cc. were used, amounting to real bleedings and reflecting technical difficulties. Bernard realized that the standardization of procedures was a guarantee for the validity of the experimental method. One wish he expressed, that will probably never come true, is that sources of error be published for didactic purposes.

The systematic search for the determining conditions of manifestations in the living, the use of the most appropriate means for controlling them experimentally, differentiated Bernard's method from that of Magendie, who represented the most advanced phase in the development of the experimental method before Bernard. Magendie's great limitation was his empiricism, strict adherence to facts with no attempt to go beyond them, and satisfaction with whatever results he obtained, even when they were contradictory. Hypothesis was rejected from fear of systems, of which there were already too many. His saying is well known: "When I experiment I only have eyes and ears, no brain whatever."[26] As a result, experiments were unplanned, and experimental conditions were not carefully analyzed. Claude Bernard converted Magendie's experimental method from empirical to scientific when it became apparent to him not only that phenomena in the living are determined in their causation but also that a strict method will reveal their determinism. As already stated, the shift was the result of laboratory investigation, of a half-missed experiment, of a hide and seek game between the investigator and the phenomenon under investigation.

According to Bernard's student, Dastre, the starting point was the puzzle of recurrent sensitivity, and Bernard's insistence on retelling the story in his courses makes it more than plausible. The story began while Bernard was nine years old and was solved by him twenty-five years later. In 1822, Magendie demonstrated experimentally the functions of the dorsal and ventral spinal nerve roots, the former being sensory and the latter motor. This is known as the Bell-Magendie law. In 1839, he discovered that the anterior motor roots are also sensory as a result of a nervous branch originating in the posterior root and accompanying the anterior nerve. This phenomenon is known as recurrent sensitivity. In 1840, the phenomenon so easily demonstrated the year before vanished, and all attempts to reproduce it by Magendie, by Longet, his scientific opponent, and by Bernard failed. It was not before 1846–1847 that the mystery was solved by Bernard—not by Magendie, who did not care to search for the cause of the discrepancy. Bernard reviewed the successful experiments from all these years. He noticed that the successful experiments had been class demonstrations performed when a good hour's time had elapsed after the surgical preparation of the animal; the unsuccessful experiments were done in the research laboratory when the nerve was stimulated immediately. Little cause and great effect—the success or failure of the experiment depended on this particular detail. The

surgical shock inhibited the sensitivity of the nerve, which became
manifest again after one hour of recovery. How Bernard varied the
experiment by progressive opening of the vertebral canal, how he
found in the facial nerve an accurate means of controlling the experiment, how he found that the hunting dog is a more suitable subject
than the shepherd's dog (not forgetting the importance of the time
element), is a beautiful laboratory adventure showing a kaleido-
scopic view of a great number of details both manual and intellectual,
each one indispensable for a successful discovery.[27] It was by putting
them together on the assembly line of mental synthesis that the experimental method was constructed. Today, recurrent sensitivity is
hardly mentioned in textbooks of physiology, but its importance to
the historian of science remains perennial. The lesson this adventure
taught Bernard is that every phenomenon is determined and can be
isolated from the intricate complexity of the functioning organism.
Determinism becomes the cornerstone of experimentation: it assesses
the existence of definite causes in the manifestations of phenomena,
claims the immutability of the relationship between cause and effect,
and thus guarantees that a phenomenon is foreseeable and repeatable
as long as the conditions of its manifestation remain unaltered. Recognition of all the circumstances surrounding a manifestation of a
phenomenon defines determinism. It explains that there are no
"poor" experiments; all experiments are good within the frame of
their determining causes. It rejects the metaphysical first cause in the
favor of the efficient or immediate causes, which are of a physical–
chemical nature and as such can be mastered by the experimenter. It
eliminates contradictions and negates exceptions. It is the only absolute law in nature because it is beyond the reach of Cartesian doubt;
the possibility of uncertainty does not arise in its case. Bernard is
probably unique in including and underlining the term "determinism" in his definition of physiology,[28] whereas the famous treatises of
Magendie, Johannes Müller, and Carl Ludwig omitted it. The term
"determinism" was coined by Claude Bernard and confused with
fatalism by those who saw in it a menace to free will.[29]

Historians are unanimous in blaming Claude Bernard for his rejection of statistics in physiological investigation, thus depriving himself of a valuable methodological tool (Faure, Henderson, Olmsted,

Canguilhem, Bernard Cohen). Unfortunately, these historians have not seen the problem in its twofold aspect of history and of determinism. There was no statistical method in Bernard's time; it should not be confused with the numerical method used by clinicians who added up cases, establishing means and percentages. Statistics was not considered as a science by statisticians themselves (Quetelet, Engel) who were still in search of appropriate procedures, of ways of expressing results, and even of a generally accepted definition. Furthermore, in 1867 at the Congress of Florence they specifically stated that statistics should "not try to invade the field of natural science."[30] No scientist used the statistical approach, and no one objects that Darwin or Wallace did not use it, although their work appears more suitable than experimental research for its applications. Bernard's objection was its incompatibility with determinism. Statistics does not reveal the underlying cause nor its mechanism of action; by alleging the existence of exceptions, it leaves the door open to indeterminism, disregarding the fact that exceptions should have their own determining causes. Laboratory work is exclusive of chance. As a procedure, statistics is of universal application, whereas experimentation requires the use of specific means for the solution of each particular problem. Statistics is true in general and wrong in particular. Without denying its limited usefulness, Bernard considered that his age should concentrate on qualitative physiology; too much remained to be discovered before a more complete view of the functioning organism could be obtained. Bernard's reply to the mathematician Joseph Bertrand should be quoted at this point: "This dog knows more about integrals than you do."[31] That is, the physiological conditions of the animal escape the impact of mathematics. Science looks for certainties, not for probabilities. The antagonistic attitude to statistics should not be confused with the mathematical expression of the obtained results, the determinism of which was already established.

To the immutability of determinism responds the universality of the experimental method, which accomplished the wants of determinism. Of universal application in its principles, it can also be adapted to each problem in particular.

The following diagram attempts to illustrate the conditions of the experimental method before and after Bernard's codification.

Figure 1

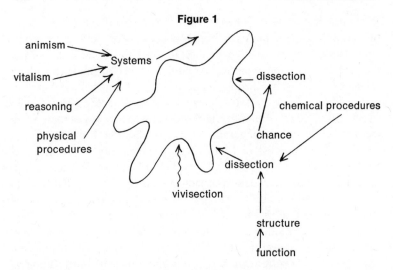

Experimental method is poorly delineated. The constituent elements are chaotically distributed.

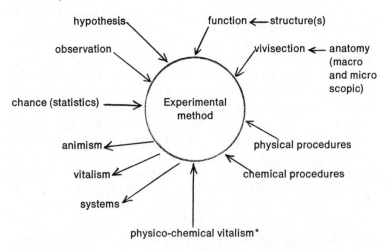

Experimental method, well delineated, polarizes component elements, rejects intruders.

* Physico-chemical vitalism means physical and chemical reactions within the frame of the organism that uses its own identifiable tools.

The second diagram represents the structure of Bernard's experimental method, the relationship and sequence of its constituent parts and the logic of the experimenter.

Figure 2

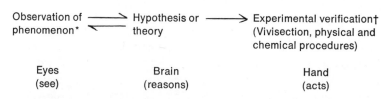

Observation of ——→ Hypothesis or ——→ Experimental verification†
phenomenon* ◀——— theory   (Vivisection, physical and
                                        chemical procedures)

Eyes   Brain   Hand
(see)   (reasons)   (acts)

* Observation may be fortuitous or purposeful (to see). At this stage the experimenter proceeds like the naturalist.
† The phenomenon is observed in the new conditions created by the experimenter; it validates or invalidates the hypothesis.[32]

At this point we may consider the program of this conference. Claude Bernard is placed between Schleiden and Darwin, and this offers a good opportunity for detecting similarities and differences between the two, representing in fact those existing between the natural and the experimental sciences, a problem that confronted Bernard. The first step in experimental conduct is identical to that of the naturalist. Whether Schleiden looked through the microscope or Darwin looked at either the beaks of finches in the Galapagos or at insectivorous plants devouring their prey, or whether Bernard watched the whitish color of the lacteals, the three proceeded by observation. This identity was at the origin of a controversy concerning the meaning of observation in the natural and experimental sciences. The objection was raised against Bernard that he opposed the observational sciences, which are passive, to the experimental sciences, which are active, and conquerors of nature. The debate, which took place at the Academy of Sciences, was led by the chemist Chevreul, the zoologist Coste and the geologist Daubrée; it was later renewed by the zoologist Lacaze Duthiers backed by Carl Vogt. They all considered that the differences between the two categories of sciences vanish as all sciences, zoology, astronomy, geology, become experimental, and that the separation made by Bernard is artificial. Chevreul was a great authority and the promoter of a *méthode a posteriori expérimentale* that had surface analogies with Bernard's experimental method. He did not appear very successful in the controversy but he compelled Bernard to define in precise terms the meaning of observation in physiology. It is true that Bernard separated observational sciences, which are contemplative, from experimental sciences, which imply the action of the scientist. He did not separate observation from experiment, however; the two are inseparable. There are two observations in the experimental method, one at the beginning and one at the end. In between is situated the action of the experimenter, who creates the second observation. Man is a creator of phenomena, said Bernard. The significance of the second observation lies in its comparison with the first; the difference between them indicates the modifying cause. The first observation shows a fact; by another fact, the second observation controls the validity of the idea that the first fact elicited. Facts alone have reality and lead to the proximal cause. The difference beween the two is that the observed fact is the product of natural conditions while the fact resulting from experimentation is the product of conditions created by the experimenter. The second fact intervenes when the first has exhausted its possibilities. Both bring information; the conditions of obtaining them differ. The unifying link between observation and experiment is their common aim, the collection of facts. At this point, Bernard met with opposition again and felt compelled to define "physiological fact," which had appeared to be self-evident so far. Through his writings and from the tribune of the Academy of Sciences, Chevreul diffused a notion of fact that Bernard found inaccurate in physiology. In Chevreul's contention, matter is unknowable as such; it is knowable by its properties, which gain existence by the interpretation of the senses that record them. In order for a fact to be known, it should be abstracted from the sum total of the properties of the organism; and since abstraction is a mental operation, it results that the concrete is accessible to knowledge by abstraction. Facts are abstractions. Bernard was at a loss in trying to understand Chevreul's definition of fact (and so am I in trying to explain it here). Chevreul's definition was unacceptable for two reasons. A fact, Bernard maintained, is whatever is recorded by the senses, either matter or its properties. For instance, the urine of a diabetic, if rich enough in sugar content, leaves crystals in the container after evaporation; this is matter. If the quantity of sugar is too small, its presence may be detected by a chemical reaction. The resulting color does not represent sugar but its property to combine with the chemical reagent. In both cases, the physiologist deals with facts. "Thus in physiology we call *fact* the material phenomenon, any act, mechanical, physical or chemical reaction."[33]

The second argument against Chevreul is that a property can be isolated from the complexity of the organism and brought to its simplest presentation without a process of abstraction. A recording apparatus, for instance, isolates and makes the observation even more objective than it is when recorded by the senses. Furthermore, it is imperative that the mind not participate in the establishment of facts which should remain free of subjective contamination; for the participation of the mind would menace the cohesion of the experimental method. The experimenter is a photographer of nature. The initial observation is and should be empirical, and should be recorded with no preconceived ideas, because the second step in the experimental method, the hypothesis, is raised by it. The mutual observation represents the stable element, which does not vary, while the hypothesis and the experimental procedures it elicits vary according to their concordance or discordance with the observed fact. The mental operation, the hypothesis, must pass the test of experimentation; and, if confirmed, it renders the second observation scientific. The hypothesis is a working tool, an anticipation of the explanation, and the only *a priori* constitutent permissible in the structure of the experimental method. If not confirmed by experimentation, it should be changed until it is fully confirmed. It should not be forgotten that much too often in Bernard's time facts had to fit the hypothesis instead of hypothesis fitting the facts. For instance, many facts were adjusted to fit the combustion theory with heat production in the lungs, a fallacious theory. An hypothesis may also originate from an already existing theory, and its confirmation by experimentation enlarges the theory by the inclusion of new facts. But it may also prove to be false and still lead to the discovery of a new fact. Examples are not wanting, and the most illustrative is the discovery of glycogen synthesis by the liver. Bernard's starting point was the erroneous theory of the day that the animal organism is unable to synthesize organic compounds, a privilege reserved to the vegetable kingdom. Searching for the site of glucose destruction in the animal, he discovered the opposite, the glycogen synthesis by the liver. It revolutionized both physiology and biochemistry.

It is essential for its validity that an hypothesis not contradict the the fact it explains, but it is even more important that it be conducive to experimentation. There are other examples in Bernard's scientific life of hypotheses leading to the discovery of new facts that invalidate the initial hypotheses.[34] The negation of an hypothesis by a fact expected to confirm it lies at the origin of another important discovery made by Claude Bernard, which opened new fields to physiology and pathology. It concerns the vasomotor nerves. Starting from clinical observations, Bernard hypothesized that the action of the sympathetic nerves is related to heat and that their severance should result in the lowering of the temperature of the innervated parts. The underlying assumption was that the sympathetic nerves regulate the nutritional changes of the innervated parts with the production of heat. The severance resulted in the opposite of the postulated theory, which was thus invalidated by the new discovery. The hypothesis had been useful, however, since it led to experimentation. The significance of this experiment is crucial for the understanding of the role of hypothesis in the method. This same experiment had been performed by numerous physiologists repeatedly since 1727 without their noticing the caloric effect. Nor did Bernard, who, like his predecessors, focused his attention on the concomitant effect on the pupil, notice the caloric effect. It was only after formulating the (wrong) hypothesis on the cooling effect that he discovered the caloric effect; then everybody saw it. Bernard centered the definition of the experimental method around hypothesis. It is "a reasoning that we use to submit our ideas methodically to the test of facts."[35]

The third step in the method is experimentation. It has two particular problems to solve: to localize the phenomenon by way of vivisection, and to explain its nature by the use of physical and chemical procedures. At this phase the experimenter becomes impersonal, a mere executant, until the end of the experiment, when he becomes again first an observer and afterwards interpreter. The manual worker replaces the intellectual, but not before he chooses the most appropriate tools from among the numerous ones offered by his arsenal of instruments. It requires discernment; the experimenter should not be slave of the instrument—a direct allusion to the German physiologists. Bernard insisted that a skilled hand is deficient if unaccompanied by a "skilled mind."[36] One should aim at "the precision of the dog, not of the instrument,"[37] the choice of the animal being the decisive factor.

The manual tools are vivisection, and physical and chemical procedures. Vivisection is the main device. It is the procedure specifically physiological and different in this respect from those borrowed from physics and chemistry. It is also the oldest; and though it had been used for two millennia and was well advanced, it had shortcomings. The leading idea of vivisectors was that each organ performs one definite function; the aim was to identify this localization. This was the anatomical conception of function fought by Magendie, whose work Bernard brought to its conclusion. The new turn was to start with the phenomenon and follow it up to the organ or organs, if any, where it was localized. Such a procedure changed the perspective of phenomena that awaited explanation. For example, nutrition is a function of the whole organism involving no specialized organ. Some physiologists denied that it is a function because there is no specialized organ to perform it. Where was digestion to be localized? In the stomach as believed for so long, in the duodenum, or perhaps in the jejunum? And what is the localized function of the pancreas and the liver? Speculation about heat production wandered from the heart to the liver, from the lung to the blood in search of a private location, until it became apparent that heat is the overall product of the activity of the organism. The same problem appeared with osmosis, a general property of all tissues.

New problems arose from the newly born science of histology and the identification of the cell. The vivisector was confronted with the technical difficulty of reaching those minute elements invisible to the naked eye where the basic properties are located—nutrition, respiration, irritability, growth. For cells the use of the scalpel, cutting and extirpation, are coarse procedures. In addition, there are numerous animal species, deprived of specialized organs, which still perform the basic functions. The originality of Claude Bernard lay in his utilization of chemical compounds with selective affinities for specific structures, thus realizing a chemical vivisection without mutilation. Examples are poisons like curare, which suppresses selectively the activity of the motor nerves, or carbon monoxide, which suppresses the oxygen carrying property of hemoglobin.

Vivisection is not an end in itself. After the phenomenon has been localized, its nature remains to be explained and its operation demonstrated. This latter point is fundamental because to Bernard "life is nothing else but a mechanism"[38] and the experimenter acts on mechanisms. This definition had the advantage of being a laboratory product that resulted from the application of physico–chemical procedures. Although the expression "physico–chemical" was used by Bernard, it is too broad and needs clarification. Physics and chemistry should be dissociated from one another; the two do not answer the same kinds of questions. Applied to biology, physical means record a phenomenon (blood pressure), and provide excellent tools for exploration (galvanic stimulus), but the intimate nature of the phenomenon is chemical. This is evident to every practicing physiologist, but not to historians, and therein lies the error of comparing the French and German physiologists of the nineteenth century. Bernard consistently used chemistry in association with vivisection and insisted on the importance of their joint application.

Another misinterpretation that reflects on Bernard's methodology is the influence of the chemistry of his time on physiology. There was no biochemistry as there was no systematic organic synthesis before 1860. Wöhler did not synthesize urea; he merely obtained it by a chemical reaction from complex compounds, but he did not proceed from simple to complex. Neither did he expel vitalism from biology, nor could a chemist perform in a test tube a demonstration that required the presence of the living organism.[39] The only available chemistry was inorganic, and the best chemists could do was to reduce complex organic compounds produced by and acting in organized structures to the simple constituent elements of the quaternary compounds (C, H, N, O). By doing so, however, they destroyed the specificity of the compounds together with their complexity. Such a procedure did not go very far in the explanation of functions. For instance, the figures for gas content in the venous blood of a secretory gland will not tell whether the gland is active or at rest since it will not take into account the part played by the nerves of the gland. The analysis is an example of a chemical solution to a physiological problem. The assertion that the chemist Liebig, like Dumas, profoundly influenced the physiology of his time[40] is untenable because their procedures made physiology subservient to chemistry. Physiologists (Magendie, du Bois-Reymond, Matteuci) and chemists (Berzelius, Chevreul) opposed this position. It was Bernard who delivered the death blow by demonstrating without the aid of inorganic chemistry that the animal is capable of synthesizing organic compounds, a feat which the chemists did not accomplish, and the possibility of which they even denied.

Another important point in Bernard's methodology concerns the relationship between organization and chemical reactions because of its connection with the suspicion that Bernard was a neo-vitalist. When he stated that the chemical nature of living phenomena is of a special order, he did not mean to imply that it obeys special laws different from those ruling the inorganic world. He merely implied that the reactions taking place in organized (not to be confused with organic) structures, organs, tissues and cells are preformed differently from those in a test tube. An example will place the difference in perspective. In the laboratory, sugar is obtained by hydrolizing starch with *acid*. In the digestive tube, sugar is obtained by the action of *enzymes* in an *alkaline* medium. The end result is the same, as chemical laws foresee, but the means used by the organism differ from those of the chemist. The physiologist is concerned with the procedures and the tools used by the organism, and this is what Bernard meant when he said that "physico–physical vitalism"[41] is totally unrelated to metaphysical vitalism. It concerns the tools used by the organism, not the nature of the chemical reaction, which is the same in both kingdoms.

The application of vivisection and physico–chemical procedures, confirming the validity of the hypothesis raised by the initial observation, does not end the task of the investigator, because the validity of the hypothesis does not guarantee the solidity of the method. For instance, finding that the content of chemical elements in the air inhaled accounts mathematically for the content of CO_2 and H_2O in that exhaled was the lucky result of a fortunate experiment. The investigator should not be subject to such coincidences. Scientific doubt must pervade the mind of the experimenter at every step of the method. He doubts the correctness of the initial observation, the validity of the hypothesis, and the proper choice of the experimental procedures. There is one thing he does not doubt, the infallibility of determinism. His way of demonstration is the counterproof, the control experiment, the safeguard of the method. It is the *experimentum crucis,* the only way to answer "yes" or "no" to the question raised by the observed fact. Everything should be put to work to demolish the confirmed hypothesis by suppressing the apparent cause of the phenomenon. If the sugar present in the hepatic veins is alimentary, as is apparent from the results obtained by Bernard and by his predecessors, a diet deprived of glucose should suppress it. The control experiment, a sugar-free protein diet, does not suppress it, thus indicating the existence of an endogenous source of glucose production, and the need to formulate a new hypothesis and to devise a new experimental approach. The initial observation alone remains unaltered. The counterproof shows the determinism of the phenomenon.

A diagrammatic presentation of one of Bernard's laboratory adventures is more eloquent than words.

Observed fact:
CO is toxic.

Phenomenon:
Venous blood (normally black) is shining red like arterial blood.

Hypothesis (1):
The venous blood in question is hyperoxygenated since its color is the same as arterial.

Experiment (1):
The blood is treated with hydrogen which normally displaces oxygen.

Result (1):
No oxygen is displaced. Hypothesis (1) is invalidated by Experiment (1). A new question is raised: if oxygen was not present to be displaced, where had it disappeared?

Hypothesis (2):
Oxygen was displaced by another gas

Experiment (2):
Blood is treated *in vitro* with CO in confined air.

Result (2):
CO in blood is not
displaced.

Conclusion: Immediate cause of CO toxicity is the suppression of O carrying capacity of red cells. The shining red color of the venous blood in question is due to the presence of CO instead of O. O is normally exchanged for CO_2 causing the change of color from shining red to black. The shining red venous blood has not been able to pick up CO_2. CO, fixed and not able to be displaced, offers the counterproof.

The following diagram represents the consequences of the most revolutionary discovery in nineteenth-century physiology, hepatic glucogenesis, for which Bernard had no precursors. It was the triumph of the experimental method.

Figure 3

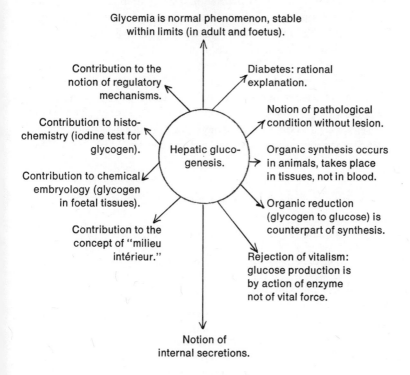

Glycemia is normal phenomenon, stable within limits (in adult and foetus).

Contribution to the notion of regulatory mechanisms.

Diabetes: rational explanation.

Notion of pathological condition without lesion.

Contribution to histochemistry (iodine test for glycogen).

Hepatic glucogenesis.

Organic synthesis occurs in animals, takes place in tissues, not in blood.

Contribution to chemical embryology (glycogen in foetal tissues).

Organic reduction (glycogen to glucose) is counterpart of synthesis.

Contribution to the concept of "milieu intérieur."

Rejection of vitalism: glucose production is by action of enzyme not of vital force.

Notion of internal secretions.

The carbon monoxide example is typical of the application of the experimental method and is interchangeable with any other example chosen from present laboratory practice. Consciously or unconsciously, the experimental method is applied by every physiologist. The guarantee of the success of his enterprise is the infallibility of its application.

The experimental method rests on "immutable foundations" because it demonstrates that there is no effect without a cause and that the cause is material. Its objectivity is guaranteed by rendering the experimenter impersonal; the validity of his own contribution is subjected to the continuous control of facts. The "great principle" is that biology "should not be studied synthetically but always analytically";[42] each single detail becomes the master piece in its turn. In this way a phenomenon "stuffed with so many truths" is decomposed, and each component part is presented in its simplicity. From there, general principles may follow the particular fact, and laws may be established. The known laws become a source of new discoveries, and the circle formed by induction and deduction is closed.

The conclusion for the historian is not so simple. In his program for experimental medicine, Bernard states: "In any case the analytic and synthetic work never ends, because the problems of nature are inexhaustible."[43] The same is true of Bernard's writings. They should be read over and over again. They seldom read twice the same way, and two readings do not contradict one another.

NOTES

1. C. Bernard, *Principes de médecine expérimentale* (Paris: P.U.F. 1947), p. 26, n. 1.
2. F. Dagognet, "Ambiguités de Claude Bernard," *Atomes* 20 (1965):351–7.
3. G. Canguilhem, *Le Normal et le pathologique* (Paris: P.U.F., 1966), ch. III.
4. Cf. comments by Grmek and Canguilhem, "Georges Cuvier Journées d'études," *Revue d'histoire des sciences* 33 (1970):32.
5. O. Temkin, "Materialism in French and German Physiology of the Early Nineteenth Century," *Bull. Hist. of Med.* 20 (1946):322–327.

6. E. Mendelsohn, "The Biological Sciences in the 19th Century," *History of Science* 3 (1964):49–50; "Physical Models and Physiological Concepts," *British J. of the Hist. of Sci.* 2 (1965):217.

7. H. Bergson, *La Pensée et le mouvant,* 27th ed. (Paris: P.U.F., 1950), p. 229. "Ce que la philosophie doit avant tout à Claude Bernard, c'est la théorie de la méthode expérimentale."

8. E. Chauffard, *De la philosophie dite positive dans ses rapports avec la médecine* (Paris: Chaverot, Leclerc, 1863), p. 25.

9. E. Chauffard, "Claude Bernard," *Revue des Deux-Mondes,* (Nov. 15, 1878): p. 40 .

10. C. Bernard, *Philosophie,* Manuscrit inédit . . . présenté par Jacques Chevalier (Paris: Boivin, 1938), p. 45. "Cette métaphysique ne se surajoute pas à sa science, elle l'inspire."

11. C. Bernard, *Introduction à l'étude de la médecine expérimentale* (Genève: Les Editions du Cheval Ailé, 1945), p. 112.

12. C. Bernard, *Physiologie opératoire* (Paris: J.-B. Baillière et Fils, 1879), p. 55.

13. C. Bernard, *Phénomènes de la vie communs aux animaux et aux végétaux* (Paris: J.-B. Baillière, 1878), pp. 50–51. *Introduction,* p. 195.

14. G. Canguilhem, *L'idée de médecine expérimentale selon Claude Bernard* (Paris: Palais de la Découverte, 1965), Série D. No. 101, pp. 14–15.

15. C. Bernard, *Physiologie générale* (Paris: Hachette, 1872), p. 10.

16. The term "metabolic force" was coined by Schwann (*Mikroskop. Untersuch.* [Berlin, 1839], p. 234). He was preceded by Gmelin who used the term "Stoffwechsel" the year before. (R. Wagner, *Handwörterbuch der Physiologie* [Braunschweig: F. Vieweg & Sohn, 1842], 1:LIII.) There is no common measure between Schwann's hypothetical idea and Bernard's demonstration.

17. C. Bernard, *Chaleur animale* (Paris: J.-B. Baillière, 1876), p. 5.

18. C. Bernard, *Substances toxiques et médicamenteuses* (Paris: J.-B. Baillière, 1857), pp. 1–16.

19. Ibid., p. 9. *Physiologie opératoire,* p. 60.

20. P. Bert, in C. Bernard, *La Science expérimentale,* 6th ed. (Paris: J.-B. Baillière, 1918), p. 19.

21. At this point, the influence of Auguste Comte should be considered. The meaning of positivism has been broadened to include literature, ethics, philosophy, sociology and science. It has become the standard of reference for all French cultural activities for the greatest part of the nineteenth century. In literature, for instance, it includes the Parnassian poets as well as Sully Prudhomme, who received the first Nobel Prize in 1901 for "lofty idealism."
Bernard's debt to Comte is still debated (D. G. Charlton, *Positivist*

Thought in France, [Oxford: Clarendon Press, 1959], Ch. V; L. Laudan, "Theories of Scientific Method from Plato to Mach," *History of Science* 7 [1968]:34), while in 1913 *La Revue positiviste internationale* sadly recognized that French authors ignore such a debt. Comte's influence on science is an immutable dogma, while the influence of science on Comte is disregarded. In the particular case of the life sciences, the influence of Bichat, Cuvier, Broussais and Blainville on Comte never comes under discussion. No mention is made of his misunderstanding of the meaning of biology, a term which he popularized and which was remote from the term of Lamarck, whom he quoted ever so often. The idea of evolution was foreign to him, as he believed in the fixity of species in a true Cuvierian manner. The absence of a discussion of the work of Magendie and Edwards in connection with the importance he attached to the *milieux* is noteworthy.

In the particular case of Claude Bernard, the distinction should be made between the historian and the experimenter. Bernard was well acquainted with Comte's work (*Philosophie,* [Paris: Boivin, 1938], pp. 25–43); he has been even accused of plagiarism. As an historian, he was definitely inspired by Comte; Whewell was apparently unknown to him. His division of the history of medicine into three stages, theological, empirical, and scientific, corresponds to Comte's theological, metaphysical and positivistic stages. The empirical stage is not as remote from Comte's metaphysical stage as it may first appear, because empiricism by its lack of causal explanation leaves the door open to vitalism.

Bernard the experimenter was not, nor could have been, influenced by Comte. Some notions were common to both—significance of final and immediate causes (a Cartesian idea), determinism, milieu (term coined by Comte), specificity of tissues, all problems that neither Comte nor Bernard invented. To Comte, they were generalities; while to Bernard, "Il faut absolument descendre dans les détails" (*Philosophie,* p. 31). Comte recommended the creation of "spécialistes des généralités" who, in Bernard's view, would have been "les êtres les plus nuisibles à toute science vraie" (*Philosophie,* pp. 30, 35). Above all, Comte denied the value of animal experimentation in general and the use of vivisection in particular (*Cours de philosophie positive,* 5th ed. [Paris: Société Positiviste, 1893], 3:250–255) and insisted on the indivisibility of the organism (Cuvier's correlation of parts). Even in chemistry, experimentation presents "de grandes difficultés fondamentales" (Ibid., 3:251). In complete opposition to Bernard, Comte declared *a priori* that experimental conditions are more suitable in lower animals, while Bernard knew from laboratory practice that the mammal is the most suitable animal for physiological analysis, because each function is accomplished by specialized structures. Comte's ideas were not conducive either to experimentation or to a correct interpretation of dynamic phenomena. (For an objective discussion, see R. Virtanen, *Claude Bernard and his Place in the History of Ideas* [Lincoln, Nebraska: University of Nebraska Press, 1960], ch. III.)

22. X. Bichat, *Recherches physiologiques sur la vie et la mort,* 3rd ed. (Paris: Brosson, Gabon, An XIII–1805), p. 32.

23. Ibid., p. 180.

24. Ibid., p. 6.

25. See also C. Bernard, "Sur le tournoiement qui suit la lésion des pedoncules cerebelleux moyens," *C. r. Acad. Sci.* 27 (1849):52.

26. C. Bernard, *Pathologie expérimentale* (Paris: J.-B. Baillière, 1872), p. 482. "Quand j'expérimente je n'ai que des yeux et des oreilles, je n'ai point de cerveau."

27. C. Bernard, "Recherches sur les causes qui peuvent faire varier l'intensité de la sensibilité récurrente," *C. r. Acad. Sci.* 25 (1847):104–105.

28. C. Bernard, *Introduction,* pp. 148–149.

29. The term determinism was used by Leibniz, Kant, Hegel, and Feuerbach.

30. J. Schiller, *Claude Bernard et les problèmes scientifiques de son temps* (Paris: Editions du Cèdre, 1967), p. 157.

31. Quoted by L. Chauvois, *William Harvey* (Paris, Sedes, 1957), p. 220.

32. J. Schiller, "Physiology's Struggle for Independence in the First Half of the Nineteenth Century," *History of Science* 7 (1968):85.

33. C. Bernard, *Physiologie opératoire,* p. 43. "En physiologie nous appelerons donc *fait* le phénomène matériel, acte mécanique, physique, rèaction chimique, etc."

34. Cf. his investigation of sugar in the urine and liver of factures. (*Leçons de physiologie expérimentale* [Paris: J.-B. Baillière, 1855], 1:231.)

35. "Le méthode expérimentale est un raisonnement à l'aide duquel nous soumettons méthodiquement nos idées à l'expérience des *faits*" (*Introduction,* p. 43).

36. C. Bernard, Mss. Cahier Rouge (p. 16), Fonds Claude Bernard, Archives du Collège de France, Paris.

37. C. Bernard, Mss. 24. d. Fonds Claude Bernard, Archives du Collège de France, Paris.

38. C. Bernard, *Principes,* pp. xix, 289; *Introduction,* p. 236.

39. J. Schiller, "Wöhler, l'urée et le vitalisme," *Sudhoff's Archiv* 51 (1967):229.

40. F. Holmes, "Introduction" in J. Liebig, *Animal Chemistry* (New York and London: Johnson Reprint Corporation, 1964), p. vii.

41. C. Bernard, *Principes,* p. xxviii; *Phénomènes de la vie communs aux animaux et aux végétaux* 2:390.

42. C. Bernard, *Principes,* p. 246; see also p. 291; *Chaleur animale,* p. 278.

43. C. Bernard, *Principes,* p. 291. "Dans tous les cas le travail analytique et synthétique ne sera jamais clos parce que les problèmes de la nature sont inépuisables."

7 Chabry, Roux, and the Experimental Method in Nineteenth-Century Embryology[*]

FREDERICK B. CHURCHILL
Indiana University

This paper ostensibly deals with a narrow and well-delineated problem. In essence, I compare two sets of experiments which were similar in design and performed nearly simultaneously in the late 1880s. The comparison has some intrinsic merit and certainly a fascination in that one set was widely advertised in its own day and is often referred to even now; the other, though not unknown at the time, merely entered the stockpile of available embryological information and its executer has virtually vanished from the annals of science. This inequity alone stirs the historian's fancy, but my interests are much more.

The 1880s were a decade when embryology changed dramatically from a descriptive endeavor closely allied to comparative morphology to a manipulative, functionally concerned, and *in vivo* operating enterprise. A comparison of these experiments will sharpen this distinction. It is hoped that it will also show that the historian must penetrate deeper than the act of organic manipulation in order to

* An earlier and shorter version of this paper was given at a symposium on the History of Experimental Embryology in Honor of the Centenary of Hans Spemann's birth and of the Retirement of Viktor Hamburger, Washington University, St. Louis, May 29, 1969. I wish to express my gratitude to the fellows of Corpus Christi College, Cambridge, for providing me the facilities for the paper's redrafting and to the National Science Foundation for support under NSF Grant #GS 2518.

161

understand the rise and appeal of experimentation in a given science. It will be argued that these two sets of experiments were undertaken for very different methodological reasons and that their differing rationales reflected more general trends and cross currents in embryology. The two investigators involved in this study are the Frenchman Laurent Marie Chabry (1855–1893) and the German Wilhelm Roux (1850–1924).

PAPERS ON HALF-EMBRYOS

In his doctoral dissertation of 1887 Chabry announced a technique for experimentally destroying the early blastomeres of Ascidian embryos and went on to describe the subsequent abnormal development of the remaining cells.[1] In a paper published one year later Roux outlined essentially the same experiment performed on frog embryos and likewise described and illustrated the resulting development of pathological half- and even quarter-embryos.[2] Roux's paper was widely publicized and stimulated a good deal of further investigation and controversy in embryological and cytological circles. It was only one, and by no means the first, of many papers Roux wrote on experimental embryology throughout a long career which propelled him to the top rank of German biologists. He performed this particular set of experiments while serving as director of his own institute for *Entwickelungsmechanik* in what was then the Silesian city of Breslau. The work earned him a full professorship at the University of Innsbrück; by 1895, his prominence still rising, he returned to Prussia, where he became director of an anatomical institute connected with the University of Halle—a position which he held for the rest of his academic life. He became, as well, the founder and editor of the *Archiv für Entwickelungsmechanik,* the major journal for experimental morphology. By contrast Chabry's dissertation was the only major paper he wrote, although he contributed a score of brief scientific notes on a wide range of anatomical and embroyological subjects. Soon after this major work he shifted into clinical medicine; he held a few posts at provincial universities before securing a good position at the Pasteur Institute in Paris. Succumbing to tuberculosis while studying the disease, Chabry died at the age of 38. Roux, living to nearly twice that age, had the fortune to see several *Festschriften* published in his honor and to hear himself repeatedly nominated as the founding father of experimental embryology.

These lancing experiments, performed by different hands and, as I shall argue, derived from different traditions, ranked among the most interesting embryological experiments of the *fin de siècle*. Their curious results guaranteed a successful hearing and challenged further investigation and explanation. Chabry's experiments comprised only a part of a general study of Ascidian development undertaken at the marine laboratories at Concarneau. His complete monograph nearly filled the entire May–June 1887 issue of the *Journal de l'anatomie et de la physiologie*, yet only in the last few pages did Chabry describe some experimental results: half-larvae with right or left sides missing because of the destruction of the right or left blastomere; three-quarters larvae resulting from the destruction of a single blastomere of the four-celled stage, and quarter-larvae from the destruction of three out of four second cleavage blastomeres.[3] Roux demonstrated his results at the September 1887 meeting of the *Naturforscherversammlung* in Wiesbaden; a report on the convention recorded his results before his extensive monograph appeared in 1888:

Herr W. Roux (Breslau) spoke on self-differentiation of the blastomeres. After the destruction of one of the first two blastomeres of the frog's egg, he pursued the fate of the other surviving cell. This cleaved, formed a *Semimorula*, then a *Semiblastula*, a *Semigastrula lateralis* and finally an *Hemiembryo lateralis*. In the further progress of the development a regeneration often commenced in the half of the body, which up till then was missing, so that finally a normal embryo resulted.[4]

Inevitably the question of mutual influence for the experimental design arises, and although I have only a mild interest in the priority charges, it is worth setting the record as straight as possible at the outset.

In a later review of these experiments Roux succinctly stated his case.[5] In June 1885, he argued, he had sent an offprint of his, "First contribution to *Entwickelungsmechanik*," to the anatomist Georges Pouchet and in return received an acknowledgment.[6] This paper entailed descriptions of puncturing experiments, or *Anstichversuche*, performed on frogs in the springs of 1882–1884. Roux recognized that Chabry had studied naturally induced half-embryos in the fall of 1884,[7] and that by April of 1885, that is, before the arrival of Roux's reprint, he had started work on his own lancing experiments under the guidance of Pouchet.[8] Roux felt it highly probable that Pouchet would have shown his student Roux's own paper when it arrived, and yet incredibly enough Chabry never cited this work nor recognized Roux's priority. It is clear from this argument that Roux considered his initial experiments of 1882 and his paper of 1885 rather than his dramatic "half-embryo" paper of 1888 (which is here under discussion) as establishing this new approach in experimental embryology.

Chabry was already dead and Pouchet died shortly after the appearance of Roux's historical note, so no rebuttals came from the other participants involved. Nevertheless, something can be said in Chabry's defense. If only the publication priority of a specific result, i.e., half-embryos from the lancing of one of two blastomeres, is considered, Chabry had clearly beaten Roux. The latter had not even started his series of half-embryo experiments until the spawning season of 1887, a few months before the appearance of Chabry's thesis. Chabry, moreover, had published a small note about his experiments in 1886,[9] and even earlier, in 1885, had drawn attention to natural half-embryos, indicating that they possessed a key to the understanding of development.[10] As for due credit for developing a method of producing injuries, both Roux and Chabry seem to have arrived independently at piercing experiments: Roux in 1882, Chabry in 1885. A lancet, after all, was the common-sense instrument for any embryologist once he had envisioned the experimental goal. Both Roux and Chabry had been anticipated to some extent by Ernst Haeckel in 1869 and Carl Chun in 1877, both of whom had observed the development of artificially separated blastomeres.[11]

More interesting to me than the details of publication are the specific contents and unstated intents of the two papers. In fact, I intend to take up here a challenge thrown out by Oppenheimer in a recent paper on French teratology. "Why?" she asked, as she addressed the problem of motivation behind these embryological studies. She went on to answer her own query with a disclaimer: "That is perhaps a question not for medical historians, but for those who presume to probe to the depths of human imagination."[12] I will not, nor could I, plummet to those suggested depths, but I do wish to look at some intermediary motives which tell the historian about some of the subtleties involved in embarking upon an experimental science. In comparing the 1887 dissertation of Chabry with the 1888 paper of Roux, I will first examine four specific aspects of their respective works: 1) the experimental techniques, 2) the justification for the use of experimentation, 3) the employment of mechanical models, and, 4) the interpretation of the results. Secondly, I will turn back to the more general issues, tentatively broached at the outset, to suggest how the findings from the case study might contribute to an understanding of the rise of experimental embryology.

EXPERIMENTAL TECHNIQUE

When one compares the details of the two sets of experiments, the differences as well as the superficial similarities quickly emerge. Chabry was clearly a master craftsman in early micro-manipulations.[13] He even displayed a childish delight with the apparatus of his own devising. He designed a jointed arm for his microscope stage, equipped with pipette, syringe and stopcock, which allowed him to isolate under low magnification a single distorted specimen from a clutch of 500 Ascidian eggs. He rigged up a vise complete with capillary tube, aspirator, appropriate stopcocks, and a rotating mechanism—all of which allowed him to view, to turn over, and to experiment on a cleaving egg under his 300X objective. His most ingenious piece of equipment was his home-spun lancet mounted on a triggering device which regulated precisely the depth of the lancet's thrust. After manipulating the egg into the correct position in the capillary tube, and after inserting the shielded lancet to the point of attack, Chabry could destroy any desired blastomere with a maximum of accuracy and a minimum of frustration. When we consider that he was working with decorticated Ascidian eggs approximately 2mm in diameter, this was no mean accomplishment. He claimed that with a slight modification the device could be used on the smaller sea urchin eggs, as well.

The apparatus was certainly of Chabry's own invention. Compared with Camille Dareste's methods for producing artificial abnormalities through his variable techniques of incubating chick eggs; compared with the mechanical agitations, the temperature alterations, and somewhat haphazard chemical treatments of fish eggs used by August Lereboullet; compared with the thermal and electrical cauterizing experiments, again on chicks, performed by Stanislav Warynski and Hermann Fol; Chabry could rightly describe his own technique as unique.[14]

Another departure was his use of a lower chordate, which, compared with the traditionally used chicken egg, gave Chabry obvious advantages in the visualization, manipulation, and preservation of the earliest stages of cleavage. Moreover, such examinations of the beginning stages promoted speculations about the potency of individual cells. "Considered from the point of view of simple normal embryology," he argued,

the traumatism of the blastomeres constitutes a new method of anatomical research. In fact, since it required merely the killing of the initial cell of an organ in order to suppress the latter, it has become possible to determine far better than hitherto the role of different blastomeres.[15]

Chabry claimed even more. His technique offered an experimental method which produced consistent and predictable results. "I am," he boasted, "with two or three eggs practically always able to obtain any monster which nature produces with the spontaneous degeneration of the blastomeres during the time of segmentation."[16] This may have been the case, but it is also obvious from the small number of successful operations and from his own confession at the end[17] that his experimental technique was difficult and time consuming.

It is obvious from a glance that Roux's technique of piercing the frog blastomeres was far cruder. It is, however, worth noting that Roux's half-embryo experiments were only a culmination of a long series of puncturing experiments which he had begun in 1882 under the curious guise of testing whether an electrical current passed through the embryo might not interfere with the differential growth of the cells.[18] Despite its failure, this initial research showed Roux that the resilient and adaptable frog embryo was a perfect experimental subject for *Anstichversuche* or puncturing experiments. So turning his full attention to piercing cleavage cells, Roux developed his technique over the next few years. He noted that some punctured embryos continued along the lines of normal development even though they exhibited a higher mortality than his control specimens; he learned that many embryos developed abnormalities which were commonly found in nature, such as an imperfect closure of the primitive mouth or a doubling of the neural folds, and finally, he found abnormalities which were unknown in the teratological literature.

During the spawning season of 1887, Roux began a new series of *Anstichversuche*. This time he first heated the needle, and for an interesting reason. "Even after repeated puncture of a cell with a fine needle," he mused,

and in spite of considerable exovation, the cell developed normally. So, beginning on the third day, I heated the needle by holding it against a brass sphere for a heat supply, heating the sphere as necessary. In this case only a single puncture was made, but the needle was ordinarily left in the egg until an obvious light brown discoloration of the egg substance appeared in its vicinity.[19]

Only twenty percent of the undamaged blastomeres survived Roux's assault. Roux immediately doused four-fifths of these survivors in a preservative as they reached the morula and blastula stages. Upon fixation and microscopic examination, Roux found to his surprise: hemispherical morulae without segmentation cavities, hemispherical blastulae with miniature blastocoeles, and semi-gastrulae with one instead of two neural folds. Among his specimens were fascinating variations including a noteworthy late semi-gastrula with a single lateral mesoderm as well as a normal, medially located notochord.[20]

As interesting as this comparison between Chabry's and Roux's techniques may be, I feel there are more revealing questions for the historian to put to these investigators. Since their work came at a moment when embryology moved decisively away from a descriptive and toward an experimental tradition, it is worth asking how each man justified the experimental approach.

JUSTIFICATION FOR EXPERIMENTATION

One does not find a penetrating discussion on the role of experimentation with Chabry. He openly accepted a distinction between natural and artificially produced monsters and took his discussion from there. He tells us that he came to do teratological work with Ascidians quite by accident. In 1884 he was studying normal segmentation when he discovered that some parental lines consistently followed abnormal and distorted segmentation patterns. He began to study these monsters, he reports, with little enthusiasm, but soon he found that these studies were fruitful in that they sometimes touched what he called the deep questions of embryology or comparative anatomy.[21]

Chabry found that he could arrange the natural pre-blastula abnormalities into seven classes—or using Isidore Geoffroy Saint-Hillaire's term, into seven *hémitéries:* 1) deviations from the set pattern of segmentation, 2) retardation of cleavage, 3) segmentation limited to the nucleus, 4) absence of cleavage altogether, 5) fusion of the cells, 6) abnormal migration of blastomeres, and 7) death of one or some of the blastomeres.[22] He pointed out that although such abnormalities might appear insignificant at the early stages of development, their effects increased in severity as ontogeny progressed. The resultant developments were the gross monstrosities which had become the standard subject matter of the teratological literature. Important for Chabry's experimental frame of reference was the possibility of understanding the connection between the *hémitérie* of early cleavage and later distortions.

By the means of his lancet Chabry could imitate only the last of the seven *hémitéries;* i.e., death of one or more blastomeres. At the outset he recognized this limitation; however, it did not fault the experimental method as a mode of research. It would be only a matter of time and ingenuity before zoologists would devise ways of producing the other *hémitéries,* as well. Chabry, in fact, suggested that compression and kneading of the eggs might induce some of these distortions.

There was a second and less trivial limitation to his lancet experiment, which Chabry explicitly discussed and which revealed his understanding of the experimental method.[23] By producing *hémitéries,* Chabry might induce a chain of events which appeared anatomically indistinguishable from one of nature's monsters, but the embryologist must not confuse the two. "The natural and artificial monstrosities," Chabry insisted, *"form two distinct series* in which there are only a certain number of similar or perhaps identical terms."[24] The killed blastomeres pierced by Chabry and the dead and gangrenous blastomeres found in nature might ultimately produce the same gross abnormalities, but the artificially induced and the "spontaneously" arisen could only be partially compared. Both developments might in their own anomalous fashion have corresponding points as the links along two parallel chains. "Experimental teratology," on the other hand, *"addressing itself to the normal egg,* permits the study of these links, but it remains mute concerning the primary cause."[25] Only when the experimenter could induce parent Ascidians themselves to produce specific *hémitéries,* Chabry added, had he mastered the complete sequence of events.

Self-evident, yes, but a curious emphasis! Chabry was extremely conscious of this distinction between normal and artificial monsters. I find in this dichotomy Chabry's rationale for and deployment of experiments in general. Two examples cited by him elucidate this point. The occasional accident on the open sea where a "fine stylet" or a nematocyst might destroy a blastomere of a free-floating Ascidian egg might act as the equivalent of Chabry's glass lancet, yet the consequent abnormal development, Chabry argued, constituted an *artificial* monster. Or again, whether it was a sitting chicken through negligence or an experimenter through design who failed to turn over a clutch of incubating eggs, the resulting abnormalities would be artificial. (One gets the uncomfortable feeling that it is only convention which restrains Chabry from calling the errant hen an experimenter!) The standard for and value of such artificial monsters became the degree to which they mimicked natural abnormalities. One reason why Chabry was so pleased with his own experimental technique was that it presented a method for starting further back on the causal chain of ontogeny. The ultimate triumph, however, would heave in sight only when the experimenter could induce the parent Ascidian to produce its own, and hence *normal,* abnormality. Chabry's lancing experiments and all teratological experiments so far devised were simply one of a large class of events which represented "the intervention of external causes during the development of a well formed germ."[26] There was no indication in his argument that the posing of appropriate questions was part of the experimental art.[27]

It lay in the nature of Roux's personality to devote many pages to the justification and explanation of his investigative approach. He wrote many articles and introductions to articles publicizing his own program of embryological research, *Entwickelungsmechanik*. He demonstrated in these ventures a mechanistic concept of the organism which allowed for the subtleties, and regulatory behavior of large molecular compounds and complex biological structures. He called repeatedly for investigators to emulate his experimental style, and at the end of his life he drew up a considerably inflated list of experimenters who he claimed had followed his lead.[28]

Roux was an advocate of the position that all the sciences passed through a descriptive phase in their development toward an experimental discipline. He gave no indication of whence came this point of view and he showed no inclination to question its accuracy.[29] According to Roux, it was altogether proper and even desirable that embryology should follow the pattern set by physics and chemistry; just as the older sciences had earlier progressed, it was now time for embryology to become experimental. There existed, in fact, a hierarchy of methods through which embryology was now passing toward a science which gave causal explanations. At the bottom of the scale stood the descriptive method, which by the 1880s had been honed to a fine art by German embryologists in particular. Roux recognized its importance and maintained that such descriptive work determined in a definitive way the patterns of normal development against which experiments would have to be measured. Next up the scale was the comparative method, which had recently been exemplified in Francis M. Balfour's outstanding textbook. Roux argued that Balfour had even made some valuable causal deductions through the comparison of developmental types, outstanding among which was Balfour's assertion that the incomplete cleavage of telolecithal eggs was due to the obstruction of a great quantity of yolk. Roux admired the inference, yet objected to it as a fast conclusion, for "one could foster the supposition to the contrary that the initial cleavage of only a part of the yolk was functionally determined since a further cleavage at the outset would not be necessary and would perhaps even upset the course of the first developmental events."[30] Roux's point was clear; only through experiments could the actual causes be established.

In Roux's mind there existed yet another intermediate stage of investigation which he labeled the descriptive experiment (*"das deskriptive Experiment"*). He did not consider this a causal analysis proper, but saw it as an essential technique for clarifying some of the events in the developmental process. Roux's own research was filled with good examples: a case in point was his very earliest experiment on amphibian eggs. Here he forced fertilized eggs to adhere to a watchglass by preventing the gelatinous membrane from absorbing a normal amount of water; the trick gave him a perceptual advantage since the eggs could not rotate, and Roux could therefore determine the relationship between cleavage pattern and the longitudinal axis of the embryo.[31]

At the top of Roux's methodological scale stood the analytical experiment. Here was the key to what he called a causal analysis. In his "half-embryo" paper of 1888 he did not discuss the advantages or characteristics of such a method, but it is clear that this investigation was one of the paradigms of what he had in mind. The definition of an experiment taken from his dictionary of *Entwickelungsmechanik* concisely describes Roux's intent:

The experiment is the artificial production of conditions of phenomena, the artificial combination of factors, in order to see what will happen because of them and in order to gain consequently a clarification of their influence.[32]

There is no attempt here on the part of Roux to work out the explanatory power or limitations of the experiment. He wrote, not as an analytic philosopher, but as a scientist who tried to formalize what he himself had done in the laboratory. One immediately notices, however, how differently he used the notion of "artificial" than had Chabry. To Roux this concept implied an experimental design and not a random accident. The operative phrases were "in order to see" and "in order to gain," and both implied that the organism could be understood through the piecemeal alteration of its parts and a logical reconstruction from the distorted effects. This was the essence of what he meant by a *causal analysis,* an expression which became almost a shibboleth for his program.

The objective of the experiment was the proper understanding of what had been altered, and here Roux recognized that unequivocal results could arise only when the experimenter attacked a single factor. The skill of the investigator was tested in the manner of his inquiry and in the accuracy of his evaluation of the factors involved. Roux demonstrated full understanding of the analytical aspect involved:

It is an art to so frame the question and so employ our means of coercion or the experimental conditions, that nature must answer us in a clear way. Therefore it is necessary beforehand to have gained a mental insight into the events to be researched, to have already analyzed mentally the process and at least conjecturally to have dissected it into its eventual factors in order then to prepare the conditions in which, where possible only one such factor is changed.[33]

When Roux turned to his half-embryo experiment of 1888, he had no concern about classifying different series. He had no uneasiness about the comparative values of descriptive and experimental teratology. The latter gave certainty where the former did not. The art was in framing the question before one began.

THE EMPLOYMENT OF MECHANICAL METAPHORS

A natural question for the historian to put to the biologist concerns the extent to which his concept of the organism determines his chosen method of investigation. Whatever the rationalization may be, the choice is bound to be more complex than would appear at first glance. There is little doubt that both Chabry and Roux employed mechanical models and in ways which could be construed as elucidating their lancing experiments.

In writing about the polyhedral shape of epithelial cells, Chabry talked about external pressures and resorted to the metaphor of grapes in a wine press. He noted that as he experimentally killed one or more of the blastomeres the survivors acquired new shapes in reaction to the changes in pressures. He remarked on the elasticity of the walls of the cells and argued that reciprocal attractions between the cells were also responsible for their shape.[34] He concluded at the end of his monograph that:

Each segmented egg, normal or abnormal, *is a system in equilibrium* and it is impossible to alter the position or the form of any of its parts without the others spontaneously and immediately assuming another state of equilibrium, just as a pile of cannon balls collapses if some one deranges one of those at the base. It is this that I have confirmed through numerous experiments.[35]

Here is a potential model which could suggest many new experiments, yet despite its mechanical insights Chabry seemed unable to exploit it in the actual design and interpretation of his research. Instead, he devoted his efforts to the description and classification of the normal and artificial monstrosities of *Ascidia aspersa* and to devising techniques to produce them. This was characteristic of the whole tradition of French teratology.

As mentioned above, Chabry, unlike his predecessors, sought and described anomalies at the cleavage stages. These early distortions focused his attention on the inherent problems of classification, for he began to calculate the combinations and permutations of *hémitéries* and realized that there existed a staggeringly large number of possibilities.[36] He pointed out that in theory the number of different monsters was equal to the number of types of blastomere alternatives (the normal state plus the possible *hémitéries*) raised to a power equivalent to the number of initial cells involved. That is: $n = a^c$, where n is the number of monsters, a is the blastomere alternatives, and c is the number of cells present in the cleaving egg at the moment of the distortion. This relationship meant that if Chabry began experimenting with the eight-celled stage, the possible combinations would be 254. To make his point more impressive, Chabry further calculated that if he could experimentally cause three types of *hémitéries,* thus allowing each blastomere four alternatives, there would exist 4^8 or 65,536 combinations—all but one resulting in abnormalities. When one remembers that a slight distortion at third cleavage became magnified in later development, Chabry's calculations implied an incredible number of monstrosities. Should each and every one be considered a distinct kind of abnormality and hence deserve a special status in a classificatory scheme? "Contrary to expectation," remarked Chabry, "the observation has shown me that far from furnishing a base to taxonomy teratogeny has demonstrated its impossibility."[37] Clearly experimental teratology had to find some criteria for classification or risk becoming mired in a swamp of empirical particulars.

Chabry suggested three sets of criteria and gave them the following ranking: 1) the seven *hémitéries,* 2) the stage of development at which the distortion appeared, and 3) the number of cells affected. Such criteria could be applied to both natural and artificial monsters; so for his experimental purposes these offered a useful guide. Yet Chabry recognized the arbitrary nature of this arrangement; he could reverse the order or find other criteria thus generating other classifications—any combination would be equally valid and equally natural:

All of these seem to me equivalent [he lamented], and none of them will avoid repeating or extending the species which are closely related on certain sides. The impossibility of classifying the monsters is a fact about which the most obstinate classifier must make up his mind, and it would be easy for me if I wanted to construct the critique here to show that the different families, classes and genera of teratology rest only on the arbitrary value attributed to these or those characters to the unjustified neglect of others. [38]

Chabry vividly saw the limits of the whole French tradition of teratology because he pursued the appearance of abnormalities back to the early cleavages in the egg. It was a triumph for his microscopy and his experimental set-up that he questioned the value of classifying distortions and their causes as a goal unto itself. Yet despite his ability to fashion mechanical metaphors about the blastomeres he was destroying, Chabry seemed insensitive to the functional questions such metaphors could promote. Those same questions, however, dawned in other laboratories.

Roux's announced program of *Entwickelungsmechanik* was in itself a testimony for his concern for mechanical causes in ontogeny. In the last analysis this appellation implied a search into the motions of development, yet there is nothing clearer in Roux's writings than his insistence that the study of the embryo should be the study of all physical, chemical and biological phenomena relevant thereto.[39] There is no nicer indication of his mechanistic concern than his description of his first *Anstichversuche*.

For the first time in the early part of 1882, and not without a secret apprehension, I sunk the point of the preparation needle into the egg which was beginning its cleavage; and thereupon entered a new manner of research which promised us a clarification of many important questions, which had been searched for in vain in other ways. I was fully conscious of the crudeness of this attack on the secret workshop of all the forces of life and compared this act to the insertion of a bomb into a newly founded factory, perhaps into a textile work, with the undertaken purpose of making a conclusion about the factory's inner organization from the change of production and its further development after the prepared destruction.[40]

The metaphor reminds one of Chabry's cannon balls, but it is worth pointing out that it implies an inquiry into the internal workings of the cells rather than a concern about their special arrangement.[41] Roux's intent was very different from Chabry's.

Entwickelungsmechanik specified more than just the appropriate subject matter; it was a method of research. Furthermore, as suggested in the last section, such a method, which Roux called causal analysis, consisted of more than the employment of an experiment, for it entailed the prior mental analysis of development into factors, which could be isolated by the experimenter. When Roux threw his "bomb" into the egg, he assumed that the disruptions would help him mentally reconstruct the normal mechanisms within the cell. As he started the new series of experiments with his heated needle, his prior mental analysis had advanced. This time he argued that the blastomeres either mutually influenced each other in development or did not. As he killed one of the first two blastomeres or one or two of the first four, Roux could test (unambiguously, he incorrectly thought) whether these cells developed dependently or independently. This alternative became one of the hallmarks of his program for embryology. "This present study," he emphasized in 1889, "is therefore only one installment, as it were, of the theme treated, that of self-differentiation."[42] When Roux set out to kill one of the early blastomeres, he was "coercing" nature to answer him "in a clear way" about a functional problem.

INTERPRETATION OF THE RESULTS

The final point of comparison concerns the pattern of development which Chabry and Roux felt they had demonstrated with their experiments. As mechanists, both were interested in the structure of the egg and its relation to the developing embryo.

Through the elimination of certain cells and through reciprocal eliminations of the other cells in control embryos, Chabry felt he could determine that the various blastomeres contained the rudiments of specific organs.[43] Thus at the four-celled stage he killed the right two, the left two, the anterior two, the posterior two, the diagonals and combinations of one and three blastomeres. He concluded to his own satisfaction that both blastomeres at first cleavage contributed to the notochord and that the left blastomere contained a rudimentary eye spot which only expressed itself when he had destroyed the right blastomere. He determined that the organs of fixation existed in a rudimentary state in the two anterior blastomeres of the four-celled stage. The eye spot and the otocyst he figured were descended from the right anterior and right posterior blastomere respectively. It so happened that at the end Chabry was unwilling to make an unqualified statement about the predetermined location of any structure, but this was a conclusion drawn from his experimental results, not a precasted commitment. He was adamant in insisting that his work in no way supported the doctrine of preformation: "Although many people will want to see in my results the decisive proof that the animal is preformed in the egg and that each part of the animal is preformed in the egg, I absolutely abhor this conclusion [je tiens à éloigner cette conclusion trop absolue]."[44] With the aid of his lancet, nevertheless, Chabry could hope to localize presumptive areas with a surety which was not characteristic of studies in descriptive embryology or comparative anatomy:

. . . but the study of normal facts, as fortunately chosen as they may be, carries less conviction than those experimental facts which can be varied in numerous manners. All the experiments [expériences], all the mutilations which one makes the egg undergo, contribute in effect to unveiling its structure, and this is certainly one of the most beautiful investigations that the naturalist is able to propose.[45]

Related to this concern over presumptive areas was Chabry's interest in cell lineage and the identification of blastomere homologies. Before his experimentation, he had been somewhat pessimistic about the success of such a quest, and his microscopic examination of segmentation patterns seemed to deny the possibility further. For example,[46] cleavage appeared in different orders in different species, and even within species the order of cleavage was not always the same. The planes of cleavages also varied greatly; at times there even existed a lack of coordination between cytoplasmic and nuclear segmentation, and finally the blastomeres were often displaced without affecting the ultimate embryo. Despite these variable phenomena, Chabry insisted that it was meaningful to homologize the cleavages if one recognized that homologous blastomeres did not necessarily render homologous organs. He lamented that he was unable to set up an experiment to test such homologies, but felt that "for the general laws of morphology" this would be important.[47] Such general concerns over cell lineage and homologies are of historical importance, for they suggest once again that the thrust of experimental teratology was anatomical.

In contrast, Roux was little concerned about cell lineage and homologizing the blastomeres.[48] During his earliest lancing experiments he had been left in a quandary concerning the role of the protoplasm in development. Sometimes the loss of much material did no permanent harm; sometimes the loss of just a fraction destroyed the developmental capacity of the blastomere altogether. By 1887 Roux was determined to discover the interaction between the protoplasm and the nucleus, which he felt must be the storehouse of hereditary material. It was with a functional question in mind that Roux directed his needle at the nucleus.

First of all, there must be substances that are not essential for development and, secondly, those whose disruption or loss in very slight quantities from the blastomere destroys its ability to develop. At the present stage of our knowledge we shall consider the latter substances preferably as nuclear components. I attempted when operating with the cold needle to disrupt the arrangements of the nuclear parts by manifold movements within the egg; as mentioned above, I was so rarely successful that I preferred to make use of heat as a destructive agent. This then accomplished the desired effect.[49]

He took pains to stain the nucleus as well as other structures, and he was careful to notice the nuclear behavior and fragmentation in the damaged half. The half-embryo experiment indeed let him conclude that the nuclear components comprised the essential substances for development.

Ever since a remarkable essay in 1883 Roux had pondered the possibility that differentiation was fundamentally a matter of qualitative nuclear division.[50] His half-embryo experiment of 1887 with the nucleus as the target gave him the opportunity to join this morphological construct with his functional concept of independent differentiation. "First of all," he decided, "it is to be deduced from the normal course of development of the undamaged blastomere that the qualitative division of the cell body and of the nuclear material, ... can proceed properly without any influence from the neighboring cells."[51] Then in an effort to clarify the meaning of this qualitative division, he proceeded to describe the developing frog embryo, "from the second cleavage on, [as] a mosaic of at least four vertical pieces developing independently."[52] Repeating his point in the following paragraph, he added that there seemed to be a "mosaic formation of at least four pieces" and suggested that only future work could establish how far along the course of embryogenesis this pattern of development continued.

The expression was not a casual choice, for in 1893 Roux devoted an entire paper to *"Mosaikarbeit."*[53] From the outset he used mosaic development as a topographical description of a cluster of independently differentiating parts. "If a whole [embryo] arises out of more or fewer independently differentiating parts, it will be composed, like a mosaic, of separately formed parts; this type of formation I have designated mosaic development."[54] Thus identified with self-differentiation, mosaic development theoretically applied to any hierarchy of units; when viewed in the larger context of the dependent–independent differentiation dichotomy, one finds no stipulation on Roux's part that mosaic development formed the exclusive pattern of development.

It is clear that Roux did not confine his interpretations to an anatomical description of the abnormal events. He wished to explain in mechanical terms how differentiation proceeded, and he combined the recent advances in cytology with his own functional queries about the interaction between blastomeres and between protoplasm and nucleus. His was a physiologically inspired experiment in that he framed hypotheses about the process of development which could be systematically tested by isolating and controlling given factors.

I have now compared four aspects of Chabry's and Roux's celebrated half-embryo experiments. There is no question that in the matter of experimental technique Chabry was the unchallenged master. Both were aware that they trod new ground when they killed the cleaving cells, but what they expected to gain from the experiments was very different. Chabry, I argued, used his experiments as an extension of descriptive and comparative embryology; he generated mechanical metaphors to describe the cellular events, but he arrived at a cul-de-sac when he discovered the arbitrariness in classifying the results. Even when he sought homologies among the blastomeres and cleavage planes, he was forced to admit only limited success. Roux, on the other hand, insisted that the experimental method secured a qualitatively different sort of information; in fact, he spent his life promoting—quite egocentrically—a program for its dissemination. He approached the embryo with functional rather than anatomical and taxonomic quests, and partly because he chose appropriate mechanical metaphors and partly because he deduced and isolated simple factors, he was able to exploit his experiments in an explicatory way. He, too, was interested in the pattern of development which his heated needle revealed, but one finds him immediately applying an explanatory model to the pattern rather than hauling out of the anatomist's closet some threadbare topic, such as the homologies of parts. It is incumbent upon the historian, however, to explore whether such marked differences had wider implications which reached beyond the idiosyncracies of these individual investigators. I have throughout the discussion already suggested that this was the case.

PERSONAL IDIOSYNCRACIES

Before I plunge into the woods to fetch out some elusive "biological traditions," it is necessary to examine how far the private lives and temperaments can explain the differences we have found.

Almost any comparison between Chabry and Roux will leave the former at a disadvantage, since the latter had ample opportunity to explain and expand upon his work. After all, did not Roux's own dissertation on the branching of blood vessels bear the style of descriptive and comparative anatomy which later characterized Chabry's dissertation? Although I shall not go into the details of his investigation, it is worth pointing out that even in his earliest papers Roux did broach some causal questions. These questions were later to become a part of his experimental method.[55] On the other hand, it is more difficult to argue that Chabry would have shown the same development in his ideas had he only lived longer. He left no hint in his work that this would have been the case, and the six years between the publication of his dissertation and his untimely death reveal no inclination on his part to pursue the matter. I feel there is more to the differences in research style than these personal equations will allow.

RESEARCH TRADITIONS

I am doubtful of an attempt to explain the differences between Chabry and Roux in terms of a psychological barrier between competing traditions of embryology. In other words, I think it would overshoot the mark to see the contrast in their research styles in the terms of "gestalt switches" which characterized Kuhn's earliest formulation of scientific paradigms.[56] Chabry and Roux would not have talked past one another in the sense of being unable to communicate. Indeed, as early as 1889 Roux gladly inscribed Chabry on the rolls of *Entwickelungsmechanik*.[57] In 1892 in a general review of pertinent literature he saw Chabry's experimental results confirming his own theory of postgeneration[58] and even mentioned that he had had an exchange of letters with Chabry about the matter. Roux simply selected out the small experimental section which concerned him, disregarded Chabry's anatomical bias, which I have discussed, and completely ignored the descriptive embryology and teratology which comprised the great bulk of the monograph. It is not that he did not understand them. Chabry, according to the context of Roux's review, did not accept the latter's notion of postgeneration, but this is hardly an indication that their different explanations of development were incommensurate. In this situation the different research styles must be understood in less dramatic terms than a psychological divide.

To the extent, however, that his recent clarifications put a greater emphasis on a community of scientists with set values, generalizations, and exemplars, Kuhn's formalization of scientific change provides a useful though hardly unique way of viewing the differences between Chabry and Roux.[59] In retrospect, the difference between the two appears to spell a division between an old approach which had reached its limits and a new effort which captured the imagination of scores of young embryologists.

FRENCH EXPERIMENTAL TERATOLOGY

I have hinted throughout my analysis that Chabry represented a French tradition of experimental teratologists; it is time to examine this tradition somewhat more closely. Oppenheimer in two essays has described the major outlines of this "school."[60] Two late nineteenth-century accounts by Camille Dareste (1822–1899) and Leo Gerlach (1851–1918), experimental teratologists themselves, cite a very similar list of predecessors and contemporary investigators.[61] Gerlach dedicated his work to Dareste; Dareste dedicated his to Etienne and Isidore Geoffroy Saint-Hilaire: "A la mémoire des deux fondateurs de la tératologie." Etienne Geoffroy's experiments of the 1820s are mentioned in all these accounts as the starting point and inspirational guide of experimental teratology even though eighteenth-century precursors are duly acknowledged. De Quatrefage, writing an essay review of the subject in the year when Chabry's dissertation appeared, believed that Isidore's major work played the formative role for the discipline; "One may say without exaggeration that his [Isidore Geoffroy] *Histoire des anomalies* did for teratology that which the *Genera plantarum* [de Jussieu] had done for botany and which the *Règne animal* [Cuvier] had done for zoology."[62] With both Geoffroys representing the starting point, and Dareste viewed as the culmination, there exists a surprisingly consistent list of practitioners: Prevost and Dumas during the 1820s, the Englishman Allen Thomson during the 1840s, August Lereboullet during the 1850s and 1860s, Stanislav Warynski and Hermann Fol from Geneva in the 1880s. Can one then garner set characteristics of experimental teratology from the major works of Isidore Geoffroy and Camille Dareste?

1) The feature which strikes the modern eye with the greatest force is the authors' concentrated efforts to describe and classify the anomalies in embryogenesis. Geoffroy devoted two and a half of his three volumes to developing a system of classification. He discussed definitions and nomenclature; he searched for appropriate classificatory schemes by means of which he could sort out the differences he found. Thus, for example, he divided the simple anomalies (*hémitéries*) into classes and orders based on the distortions in volumes, form, intimate composition, disposition of parts, and multiplication or absence of specific organs. He constructed out of the "Embranchement" of "Monstruosités" a complex hierarchy of "Classes," "Ordres," "Tribus," "Familles," and "Genres," which remind the historian of the eighteenth-century efforts of zoological and botanical systematists or of the medical nosologists.[63]

Although Dareste's classificatory emphasis was different, he accepted Geoffroy's system with only small modifications.[64] He devoted the last two of the three parts of his text (approximately two-thirds of the volume) to describing types of monstrosities in microscopic detail and to examining as nearly as possible their moment of inception. He insisted that all the types of simple and compound monsters resulted from a modification of very early embryonic life, at which time either an arrest of development or a fusion of similar parts occurred.[65] This "fundamental fact," as he called it, underscores the morphological underpinnings of his teratology, since it shows Dareste absorbed with the special arrangement of anatomical parts. Both Geoffroy and Dareste at heart were playing the roles of comparative morphology.

2) Of primary interest for my study is the extent to which Geoffroy and Dareste performed teratological experiments and the explicit goals for such manipulations. As he set out to produce artificial monsters, Geoffroy felt he was following very much in the footsteps of his father.[66] He visited the chick eggs with a great variety of torments: standing them on their ends, cutting them open, and varnishing and perforating their shells. His manipulations were to no avail since they either failed to produce anomalies or killed the embryo outright; nevertheless, Geoffroy was explicit about why he, and his father before him, undertook such efforts. It appears that both Geoffroys wished to marshal evidence against the eighteenth-century conviction in preformation—a belief which clearly must have had a following well into the nineteenth century. Birth defects caused by physical and mental disturbance of the mother during pregnancy and Isidore Geoffroy's statistical studies of anomalous births among the working classes both indicated that these certain embryonic defects must be the products of fetal disruptions rather than of preformed distortions. Geoffroy's attempt to produce more monsters experimentally was simply another way of adding to the body of teratological evidence against preformation. Experimentally produced monsters were more numerous and more convenient to study than natural monsters, but there was no indication on Geoffroy's part that they possessed an explanatory power which was qualitatively different from the other types of evidence he had gathered.

Dareste was much more systematic in his experimental work. Thousands upon thousands of experiments combined with microscopical studies allowed him to trace the anomalies back to the earliest organ-forming stages of development. What was more, Dareste argued that experimentation permitted the teratologists to find terms which were otherwise unknown, and to read generalizations therefrom. Chemistry was thriving on just such an approach: "From all the parts it [chemistry] forms the bodies which it studies, and finds in this formation the laws which govern their constitution."[67] Dareste intended to pursue the same tack with his teratology. The closing passage of the first edition of his work emphasized how the experimental method allowed him to touch upon the larger issues in biology:

But behind all these facts which I have discovered, there is another much more general fact: that is that I have myself produced the elements of my study, except for the case of the double monstrosity. By modifying the external conditions which determine normal development, I have forced the appearance of all the teratological types. Of all my results I attach to this the greatest importance because it demonstrates the present limits and future application of the experimental methods to the new popular questions of zoological morphology. It is not in accumulating some relatively convincing hypotheses nourished by long and sterile discussions that one will one day determine the origin of the form of life. If the problem is accessible to us, if it does not extend beyond the scope of human intelligence, only experimentation can furnish the solution. Following this line of thought, Etienne Geoffroy Saint-Hilaire tried to produce monsters. I have also followed in the footsteps of the great naturalist during this long series of investigations which I have just completed, and which lead me to reunite the elements from teratogeny. I have the strongest hope that this thought will be fully justified by the science of the future.[68]

It is clear from this passage that Dareste placed a greater burden on his experiments than Geoffroy, but just as the latter, Dareste employed his experiments for a descriptive rather than explanatory mission. He, no less than Geoffroy, was concerned with relating different experimental conditions to given abnormalities, and the great mass of his book entailed just such correlations. On the more general level, instead of being concerned with the hoary debate over preformation of germs, Dareste used his experiments to attack the problem of the origin of races. The two problems seem hardly related from a modern context; however, by joining his name with Etienne Geoffroy, Dareste implied a different perspective. In the last analysis Dareste, as well as both Geoffroys before him and Chabry contemporaneously, employed the experimental method in order to produce forms for anatomical and taxonomic studies.

3) Both Geoffroy and Dareste worked within the spirit of comparative morphology, which meant that they saw their own teratological work contributing to the discovery of the general laws of animal organization. This was hinted at in the above quotation.

Geoffroy was forthright about his contributions, perhaps because he saw himself laying to rest the ghosts which haunted the phenomena of monstrous births.[69] The era of fables was past; the time for simply classifying differences was also over. "Zootomists" were now turning to the study of analogies and were discovering many general laws for zoology, anatomy and physiology. "This same philosophical spirit," Geoffroy noted, "happily applied to the observation of anomalous beings, had led to the same results, and like zoology and ordinary comparative anatomy, teratology has presented its generalizations, its principles and its laws."[70] The big question which Geoffroy had to settle was whether the teratological laws thus derived from the study of abnormalities bore any relationship to the zoological laws derived from normal organisms. If he could establish such a correlation, he could maintain that teratology joined the other biological fields in a common quest for unifying principles. The details of his demonstration need not detain us.

It is sufficient to point out that Geoffroy compared the frequencies of abnormalities of different organs with the frequencies of normal variations and argued that both categories were explainable in terms of general laws of growth. He noted, as well, that abnormalities and natural variations were subject to the same laws of symmetry and affinities. By the end of his discussion he could triumphantly declare that:

The analogy between teratological and zoological laws is real and striking; when one makes a sufficiently detached comparison, it attains perfect identity. It is no longer the special zoological laws nor the teratological laws but the general laws which are applicable to all manifestations of animal organization and which, embracing as so many secondary considerations all the generalities, are limited to a single order of facts.[71]

By Dareste's generation there seemed to be no need to persuade the community of scientists that teratology was a legitimate biological study. Dareste, nevertheless, was careful to point out that the experimental production of anomalies contributed to the understanding of the great problem of the day, i.e., the transformation of species, and thus contributed to the general laws of animal organization. Dareste went so far as to assert that "My experiments therefore give to zoologists some methods which will aid them to approach scientifically the question of the formation of species."[72] It was not that he himself had created new races, but that in demonstrating the influence of the environment on embryonic development, he had pointed the way for future research. If zoologists moved from the confines of laboratories to great menageries where they could experiment on an entire sequence of generations, and if they learned how to influence gametes as well as fertilized eggs, then, Dareste argued, true species transformation would be possible. His thoughts tripped nicely to the rhapsodies of the Lamarckian revival and at a time when there reigned genuine confusion between the organization of individual animals and a genetic concept of species. For this reason Dareste had a legitimate claim on professional attention, and he was gratified that Darwin himself noted the importance of his work.[73]

Dareste also recognized that his endeavors went beyond the unifying scope of Geoffroy's work, since his investigations of chick embryos demonstrated that their classificatory schemes applied to the avian as well as mammalian class. Dareste made even further connections. By utilizing von Baer's laws of development, he reasoned that anomalies which appeared at the earliest stages of ontogeny were of the most general type; on the other hand, anomalies which showed up only later were more restricted in nature. "I have the conviction," he concluded, "that it is moreover only a question of converting in order to transform the special teratology of the chick into the general teratology of vertebrate animals."[74]

Examining thus the works of Geoffroy and Dareste, I have extracted some common strands which strike me as significant for understanding their experimental method. They had anatomical and taxonomic rather than physiological concerns; they resorted to the manipulation of the eggs not to gain a deeper understanding of their functions but to increase the variety of forms. At the core of their research lay morphological quests about the fundamental organization of living forms. Since Chabry so closely identified himself with Geoffroy, Dareste and other French teratologists, it is not surprising to have found that his research bore the same characteristics.

THE GERMAN SCENE

In turning to Roux's background, one is much harder put to identify a well-defined tradition; Roux himself, after all, claimed to be the founder rather than the disciple of a school of embryology. It is worthwhile, nevertheless, to examine his formative years, for they make his attitude toward experimentation more understandable. There were two distinct periods in Roux's early career: the first, his student years at Jena between 1871 and 1878; the second, his years in Breslau between 1879 and 1888. By the time he accepted a professorship in Innsbruck in 1888, he had defined his research goals well.

Roux wrote his dissertation at the medical faculty under the supervision of the young anatomist Gustav Albert Schwalbe (1844–1916). Schwalbe filled the professorship of anatomy at Jena between 1873 and 1881 after having worked in Eduard Pflüger's physiological laboratory in Bonn, having earned his doctorate in Berlin, and having spent two years as associate professor of histology at Carl Ludwig's laboratory in Leipzig.[75] His biographer claimed that Schwalbe developed a strong interest in the borderline concerns contingent to histology and physiology, and it is pertinent to note at least the titles of some of his papers published during his years at Jena. Thus one finds: "Beiträge zur Kenntnis des elastischen Gewebes (1876)," "Über das postembryonale Knochenwachstum (1877)," and "Über Wachstumsverschiebungen und ihren Einfluss auf die Gestaltung des Arteriensystems (1878)." At the end of his life Roux remarked that Schwalbe had very little experience with embryological matters and that he himself had to work out his own techniques.[76] This may be a perfectly fair judgment, but it should not be allowed to conceal Schwalbe's influence at a deeper level. The titles of all three of the above papers indicate that Schwalbe had a lively interest in the mechanical interaction between growth and structure. Roux's own dissertation, guided by the same concern, was an attempt to relate the pattern of arterial branching in embryonic muscles to hemodynamic forces.

Ernst Haeckel (1834–1919) was the most domineering personality on the Jena scene during Roux's student days. With the retrospect of half a century Roux claimed that although he attended some of Haeckel's lectures, he never worked closely with the persuasive and volatile zoologist.[77] Perhaps more important than a direct tutelage for broadcasting his personal style was Haeckel's major theoretical work, *Generelle Morphologie der Organismen,* which was published five years before Roux's matriculation at Jena.[78] This was one of the great compilations of the period. Its author, synthesizing the contemporary advances in anatomy, embryology, cytology and evolution theory into a grand ordering of the phenomena of life, gave vent to his monistic world picture. Haeckel, a morphologist admiring the reductionist approach in physiology, exhorted biologists to banish the last vestiges of vitalism from their studies and to accept only the physicists' and chemists' world of matter and forces. He equated teleological reasoning with all forms of vitalism; he identified causal reasoning, its opposite, with a professed commitment to the sixty known chemical elements and to an aether as the sole ontological entities. Elements and aether alike were composed of atoms endowed with the traditional mechanical properties of impenetrability and indivisibility. The chemical atoms, however, possessed a mutually attracting force while the smaller aether atoms possessed a mutually repulsing force—it was this difference which kept the universe in a state of perpetual stir. The monistic philosophy envisioned a bonding of matter and force in such a fundamental way that one could interpret all phenomena as being the necessary playing out of their initial state. Haeckel contrasted in great detail inorganic and organic forms and events, but he argued that all the differences he ultimately found, ranging the spectrum from the properties of colloids to the human psyche, could be attributed to the necessary consequence of the fundamental union of *Stoff* and *Kraft.* Haeckel went so far as to redefine God himself as the sum of all forces and all matter. "Gott ist das allgemeine Causalgesetz. . . . Gott ist die Nothwendigkeit."[79] It is evident from his early work that Roux was profoundly taken by the mechanistic analysis of life presented in the *Generelle Morphologie.*[80]

A third teacher in Jena who left a strong impression on Roux's thoughts was the young professor of physiology, Wilhelm Preyer (1842–1897). Preyer was an enthusiast for Helmholtz, Du Bois-Reymond, and Darwin, and was best known for his physiological studies on hemoglobin, color vision, sleep, and the psyche—a set of problems examined also by the German "reductionist" physiologist after mid-century.[81] Of more direct bearing on Roux was Preyer's work on the physiology of development, which finally appeared in the form of a large tome.[82] Preyer felt that physiologists, by concentrating exclusively on the adult organism, had neglected the physiology of the embryo. His text of 600 tightly printed pages was devoted to experiments and analysis of such standard physiological problems as rates and variations in the embryonic heart, embryonic waste products, gas exchanges in the fetus, nutritional variations, temperature variations, embryonic movements and reflexes, and embryonic sensations. Preyer did not try to explain differentiation, and it should be clear from the above list of problems that all he had done was to transfer the customary physiological concerns to a stage in embryonic life. Nevertheless, he subjected the embryo to experimentation and articulated more fundamental problems. "The commanding duty," he insisted at the end of his book, "confronting the physiologist before anybody else is to tackle experimentally the great problem of development and to lay the concept of heredity into its parts."[83] At the end of his life Roux stated that he had entered the medical faculty at Jena at Preyer's suggestion and had attended his lectures;[84] on occasion he explicitly recognized Preyer's influence on his own work.[85]

It is hard to portray the impact of the Jena years on Roux's intellectual development in more precise terms than these suggestive intellectual vignettes. Roux certainly was confronted with the analytic, mechanistic and physiological attitudes presented in a collective fashion by this triumvirate of instructors; it seems reasonable to suppose that when the same attitudes reappeared in his own work they drew sustenance from Roux's university heritage. Further, it is worthwhile pointing out in passing that these men were all young and on the make. When in 1874 Roux began his first research project, Haeckel was 40, Preyer 32, and Schwalbe 28—surely an added attraction for an ambitious student of 24.

In 1879 Roux became second assistant at Carl Hasse's anatomical institute at Breslau. If his student years at Jena had persuaded him to set a philosophical and functionally oriented framework to his embryological work, the decade spent in Silesia marked a time when he formulated his vague questions into specific investigations. It was the time when he began probing the frog's egg and when he eventually destroyed the nucleus with his hot needles. At the institute Roux had the association of another assistant who possessed a rather similar functional bias.

Gustav Born (1851–1900) was the prosector. A year Roux's junior, he had studied medicine at Breslau, Bonn and Berlin.[86] Born had been particularly influenced by the physiologists Rudolf Heidenhain and Eduard Pflüger, and over a span of twenty years he performed many experiments on embryological problems which were closely related to Roux's own work. When in 1882 Pflüger performed the rotation experiments on amphibian eggs which prompted Roux to investigate the influence of gravitation on cleavage patterns, Born performed his own analytic experiments on the same problem.[87] It may be impossible to untangle the mutual influence Born and Roux had on each other; it is sufficient for my purpose to recognize that Roux had a comrade in arms just as he began formalizing his experimental method.

One may say of his associations and education that Roux was exposed to a mental convention which placed a premium on analytic inquiry. His own experiments with blastomeres were in subject very different from the work he or his teachers had done at Jena. On the other hand, he had learned from them to dissect the organism mentally into its functioning components and to manipulate physically the living processes in order to understand functions rather than structures. That his colleague Born appeared equally at home with this attitude suggests that in the 1870s and 1880s in Germany the would-be embryologist was being drawn into a physiological mode. His counterpart in France, if Chabry was at all representative, went through the motions of a manipulative act but continued to think along morphological patterns. It is to the epistemological differences of these two approaches that I now turn.

EPISTEMOLOGICAL CONCERNS

There were two striking features of Geoffroy's vision of scientific progress.[88] First, he maintained that each branch of science passed through three stages of growth before attaining full maturity: a period of ignorance and superstition, a positive period of fact collection and verification, and finally a philosophical period when general laws were discovered. It is important to emphasize that in this scheme each branch of science exhibited the same developmental pattern, and Geoffroy added some detail as to when teratology had attained each stage. Second, Geoffroy insisted that all sciences maintained a palmate or radiating pattern with respect to each other. This implied that as science became more specialized it branched into disciplines which in turn branched into sub-disciplines. As this divergence continued, however, a countervailing process set in, since generalizations and laws were found to bridge the gaps between disciplines. "Thus, each science tends to fractionate, to divide for the study of detailed matters; [on the other hand] to unite and to join up for research into general matters."[89]

Now, what I find important in Geoffroy's image of science is the non-hierarchical structure of the whole. Each branch terminated at the frontiers of basic knowledge; each branch had immediate access to the general laws which bound all into a unity. Thus the teratologist was as likely as the anatomist to find laws which unified all of morphology, and by an extension of reasoning, biologists could work out the general laws of nature as well as chemists and physicists. The teratologist in no way felt that he wielded less explanatory power than the physiologist; in fact, in the 1830s Geoffroy had far from an idolatrous regard for the state of that sister discipline—"a poor and too often a conjectural science . . . possessing few facts and embracing questions in its studies which are immense in number and infinite in complexity. . . ."[90] For Geoffroy and for morphologists who adhered to such a palmate or radiating view of the history of science, the truth about nature was not to be found through a surrender to other disciplines, since that recourse simply implied an established hierarchy. Maintaining its own sovereignty, each branch contributed to the advancement of all of science.

A generation later in France the scene was different. Discussions and even debates about the differences between branches of science centered more on the method of research and less on the assigned domains of investigation, which had been Geoffroy's concern. By the 1860s and 1870s zoologists were quite abruptly confronted with spectacular gains in physiology and by some pointed challenges on the part of Claude Bernard. One detects, however, in the resulting controversies a jealous reaction on the part of zoologists to maintain their identity and one can even discern a residue of Geoffroy's concept of scientific progress and anti-hierarchial stance. Such a reaction if prevalent also renders Chabry's attitude toward experimentation understandable.

One of the most explicit expressions of this anti-hierarchial view of science came from the pen of Henri de Lacaze Duthiers (1821–1901) at the time when Lacaze Duthiers published the first volume of his *Archives de zoologie expérimentale*.[91] The event which whetted his anger and prompted this open assault was a report by Bernard on the state of physiology.[92] The report, Lacaze Duthiers felt, had unjustly and derogatorily portrayed zoology as a contemplative, and by nature non-experimental science. This charge had rankled the sensitivities of many a zoologist who had endeavored to act experimentally, and Lucaze Duthiers took the opportunity to counterattack by giving a brief review of the history of zoology and challenging directly Bernard's definition of the experimental method.

The survey of the forefathers of zoology fell into a pattern not unlike that presented by Geoffroy. Linnaeus, according to Lacaze Duthiers, had brought an end to the chaos of names and facts of an earlier phase of natural history by a reform in nomenclature, a concentration on accurate descriptions and a precise system of classification.[93] At the turn of the century Cuvier and his followers had ushered in a second reform by uniting zoology with anatomy and embryology in a search for the generalities of animal organization: "Acquainted with an understanding first of organization [they] preoccupied themselves with the general relationships of existence."[94] Zoology, Lacaze Duthiers continued, was now entering yet another phase, that of experimentation;[95] he left it, however, to the reader to draw a connection between this assertion and the title of his new journal.

In the next breath Lacaze Duthiers made it clear that in employing the expression "experimental method" he was not limiting himself to Bernard's definition. The leader of French physiology had likened the experimental style of his discipline to the methods of physics and chemistry. In short, where the latter analyzed and dissected inert matter into its parts, the former endeavored to do likewise with organisms. Where the chemist was intent upon explaining chemical reactions in terms of the chemical elements, Bernard wished to view all life in terms of histological units. In both cases, Lacaze Duthiers argued, the exclusive efforts to analyze implied a reduction to a lower order of organization. "This preoccupation is easily understood since the novelty as well as the originality of the doctrine is found in this sort of simplification of organisms reduced so to speak to some elements which need only to be studied in isolation."[96] Going one step further, Lacaze Duthiers insisted that Bernard's reductionism was a result of a particular form of determinism which upon examination turned out to be nothing more than a brand of materialism. Even though the great physiologist professed to leave his metaphysics at the door of his laboratory before entering, Lacaze Duthiers paraphrased Bernard's well-known metaphor, "One is tempted to believe that the day when his *Report* was written the door of the laboratory was left ajar."[97] The *coup de grace* of the historical review came when Lacaze Duthiers insisted that this determinism and methodological reductionism led Bernard to the restrictive concept of the experimental method which the zoologists found so obnoxious:

The interpretation given here to the meaning of the word experiment [expérience] is a necessary consequence of the goal pursued by M. Cl. Bernard, for if one abandons it, *determinism* reduces to the determination of the properties of tissues, the field of experimentation immediately extends indefinitely, and physiology is no longer able to dream of reserving experimentation exclusively for herself.[98]

In his defense of zoology Lacaze Duthiers rejected the hierarchical implications of Bernard's *Report*. Zoology was not inferior to physiology in explanatory power; it, too, could be experimental, and most important the experiments which issued from the hands of zoologists need not be restricted to the physiological mode of analysis, that is, a reduction of living phenomena to more basic levels of matter. Throughout the rest of his essay, Lacaze Duthiers gave examples of zoological experiments to prove his assertion and drew upon an alternate definition of experimentation. One illustration and the source of the definition are worth citing in order to complete this background review of Chabry's experiment.

Lacaze Duthiers related the story of how he discovered the origin of double monstrosities of the mollusk, *Phylline asperta,* and so "was able to produce the double monsters at will." He noted in the observations of the mollusk that such monsters were produced at a time when the outgoing tide interrupted the activities of a parent which was depositing its eggs in the tidal flats. Thus, compelled to hasten the ovipositing, the mollusk placed two instead of a single egg in each encasement of slime and sand. During development, the eggs fused and a double monster resulted. This chain of events once worked out made it possible for Lacaze Duthier to disturb by hand the laying adults and to create for himself the monsters which nature had produced by accident. "Was this an experiment or contemplation?" Lacaze Duthiers demanded:

Certainly there was here a provoked observation, certainly also a result was obtained in a constant and definite manner. Save, then, for direct action on a tissue element everything is gathered here which identifies an experiment. To employ the very expressions of Claude Bernard, have I not been "the experimenter who produces the phenomena of whose conditions I am the master?" Defeated on this point, one of the most brilliant exaggerators of the experimental school said to me: "But you are doing physiology here; and this is not zoology."[99]

No wonder the zoologists took offense! I might also note in passing the parallel between Lacaze Duthier's analogy of the tide to the embryologist's act on the one hand and Chabry's comparison of the floating namatocyst to his glass lancet on the other.

Lacaze Duthiers called upon the authority and prestige of the aged chemist Michel Eugène Chevreul (1786–1889) for a counter definition of the experimental method. Chevreul had often pondered this question in the light of his own work and had developed what he called the *a posteriori* experimental method.[100] Briefly, this consisted of a three step process which involved 1) the identification of a problem, 2) the isolation of the pertinent factors, and 3) the control. His examples of discovering the dyeing properties of various solutions illustrated the procedure. Thus in determining the difference of the coloring actions between the water of the Seine and the wells of the Gobelin works, Chevreul had first measured and standardized the effects on chromatic circles of different textile materials. Secondly, he had looked for the chemical cause of the diverse effects by a chemical analysis of the two waters. Thus he had found certain organic materials in the Seine and a copper salt in the Gobelin wells. Finally, "It was necessary . . . to have demonstrated in a precise manner that such compounds contained in the waters were really the causes of those determined effects."[101] This was where the crucial control came into operation which changed an empirical analysis into an experimental investigation. Chevreul had prepared solutions of the identified compounds in distilled water and had tested this against the chromatic circles. If the same effects did not occur, he had to retreat to step two and search for other compounds. Through such a controlled examination Chevreul had discovered that only a combination of calcium and copper carbonates produced the highly prized *"rougir la couleur du fustet."*

Chevreul argued that with added sophistication the same method of control could lead to an understanding of the therapeutic effects of drugs. Lacaze Duthiers found in this *méthode a posteriori expérimentale* an accurate description of his own research. One can well understand that his manual disturbance of the laying adult of *Phylline asperta* constituted the control which verified his analysis of the chain of events. By following Chevreul's procedure, his was indeed an experimental science regardless of the comments of the physiologists. It is also worth noting that Chevreul, like Geoffroy and Lacaze Duthiers, inclined toward a branching rather than a hierarchical image of the sciences.[102] Thus Chevreul's method not only gave luster to the zoologists' defensive maneuvers against Bernard but vindicated non-reductionist scientific research.

With the present evidence one can hardly prove that Chabry subscribed to a non-hierarchical view of the history or structure of science or knew of Chevreul's *a posteriori* method of experimentation. He did work for a while under the direction of Lacaze Duthiers, who was professor of anatomy, comparative physiology and zoology in the science faculty at the Sorbonne,[103] and the original line of his investigation, namely the descriptive embryology of Ascidians, makes it reasonable to suppose that he shared the interests of French zoologists. What is important, however, is that his experiments, with their orientation toward reproducing natural monsters and finding general anatomical laws, and their disregard of an analytical inquiry into the cellular forces of development, suggest that Chabry toiled in the same methodological and philosophical vineyard. Embryological knowledge was to be viewed independently of physics, chemistry, and physiology.

In turning to the epistemological basis of Roux's work, my treatment again emphasizes the personal contacts and statements rather than a tradition of attitudes. Despite this inequity in the comparison between Roux and Chabry, certain modes of thinking emerge which get to the basic differences of the two experiments.

I have already unveiled Roux's views of the historical development of embryology in the section where I examined his definition of experimentation (Justification for Experimentation). His opinion stressed a progression of techniques which started with anatomical descriptions, passed through comparative and descriptive-experimental stages, and finally reached the stage of analytical experiments. Superficially this progression of technique did not differ greatly from Lacaze Duthier's historical account, but Roux in no way suggested that the fields of science bore a branching and independent relationship to each other. To the contrary, the analytical experiment made it possible to bring physiological, chemical and physical notions into embryology; to see biological research as an attempt to reduce living phenomena to chemical and physical phenomena. Roux was too much a biologist to claim that he had indeed performed this reduction in his own work; in fact, at times he even suggested that the biologist may never accomplish the feat. But such programmatic doubts did not negate his metaphysical commitment. Causal analysis offered a beginning, and it was clear that success could only lie in the piecemeal dissection of the organism into its grosser then finer parts:

With these efforts to trace the organic developmental processes to less complex and to truly simple physico-chemical components, we must first tie in with the current biological analysis of the organism: i.e., with the dismemberment of the complicated organism into organs, the organs into tissues, the tissues into cells and intercellular substances, the cells into cell bodies, such as the nucleus, centrosome, cell membrane etc.[104]

Scores of similar statements can be found in Roux's writings. There was no question that he had an hierarchical view of the structure of science and that this was based on a strong opinion about the nature of the world. His conviction meant that the embryologist not only sought to emulate the analytic methods of chemists and physicists but tried when feasible to reduce his matter to theirs. Roux and Born interpreted the experimental and functionally oriented style of the physiologists as leading down this very path; therefore, instead of resisting their invitation (with Bernard a demand!), they joined the merry band. It was fully appropriate that Eduard Pflüger, the man who stimulated Roux's and Born's initial experiments, was a physiologist; it was symptomatic that Roux should complain at the end of his career that physiologists, such as Carl Ludwig and Adolf Fick, rather than anatomists and zoologists appreciated his earliest experimental work;[105] finally it was fitting that the physiologist Rudolf Heidenhain should suggest to Roux the appellation of *"Entwickelungsmechanik"* for his life's research.[106]

In heading this section "epistemological concerns," I wanted to stress the methods of acquiring knowledge about reality rather than the metaphysical commitments. There is no way of telling from his dissertation what Chabry's commitments actually were. His use of the experimental method to arrive at general laws suggests, however, that had he been a consistent thinker, he would have had a very different notion from Roux about the ultimate organization of things. Perhaps the best I can do is to point out that Chabry's and Roux's pursuit of morphological and physiological generalizations were significantly different and possibly opposed.

CONCLUSION

I started out by comparing two sets of experiments which were executed during a decade when embryology changed from a descriptive to an experimental endeavor. For all intents and purposes these experiments could not be distinguished in any earth-shaking way as far as the procedure and results were concerned. The results of half- and apparent half-embryos contributed to a level of debate which

included important issues, such as the mechanism of species formation and the dilemma over preformation and epigenesis. In neither
case did these experiments render an unambiguous interpretation.

On the other hand, it is clear that when considering the advent of
experimentation in nineteenth-century biology, historians must contend with more than the manipulative act and theories advanced by
the results. Both Chabry's and Roux's discussions suggest deeper
concerns which welled up from subterranean beliefs about the structure of science and the primacy of certain types of scientific data.
That their beliefs differed becomes apparent when historians contrast Chabry's spatial use of mechanical metaphors and quest for
anatomical laws with Roux's analytic employment of metaphors and
functional approach. The differences become more understandable
when we investigate the teratological tradition in France and Roux's
academic heritage. They become most explicit when we realize that
these beliefs fostered either a hostile or receptive behavior toward
the rising star of physiology.

If my analysis of the French teratologists and Roux's background
is valid, it should give historians pause for reflection. Despite our
elaborate monographs and professional standards, we are, in our own
efforts to write the history of science, as much prisoners of convention as were Geoffroy, Lacaze Duthiers, and Roux in their modest
historical sketches of their own disciplines. One cannot write the
history of science without some concept of the structure and development of science, and ultimately this concept will depend on some
theory of knowledge; too often this is an unconscious dimension to
our work. To add a parting quip: even when the historian limits himself to a simple comparison between two experiments, the very selection of items to discuss reveals his unspoken bias.

NOTES

1. Laurent Marie Chabry, "Contribution à l'embryologie normale
tératologique des ascidies simples," *Journal de l'anatomie et de la physiologie normales et pathologiques de l'homme et des animaux* 23 (1887):

167–321. For two brief obituaries of Chabry see Georges Pouchet, "Décès de M. Chabry," *Société de biologie, Comptes rendus hebdomadaires des séances et memoires* 45 (1893):919–20; and "Laurent Chabry," *Journal de l'anatomie et de la physiologie* 29 (1893):735–9. For a recent discussion of Chabry, see Jane M. Oppenheimer, "Some Diverse Backgrounds for Curt Herbst's Ideas about Embryonic Induction," *Bull. Hist. Med.* 44 (1970): 241–50.

2. Wilhelm Roux, "Beiträge zur Entwickelungsmechanik des Embryo. Nr. V. Ueber die künstliche Hervorbringung halber Embryonen durch Zerstörung einer der beiden ersten Furchungszellen, sowie über die Nachentwickelung (Postgeneration) der fehlenden Körperhälfte," *Archiv für pathologisches Anatomie und Physiologie und für klinische Medizin* 114 (1888):113–153, 246, 291. For an English translation of the major portion of this paper, see *Foundations of Experimental Embryology,* ed. Benjamin H. Willier and Jane M. Oppenheimer (Englewood Cliffs, N.J.: Prentice-Hall, 1964). For biographical information see Roux's own autobiography, "Wilhelm Roux in Halle a.S.," *Die Medizin der gegenwart in Selbstdarstellungen,* ed. L. R. Grote (Leipzig: Felix Meiner, 1923), 1:141–206. For recent appraisals of Roux's work, see various papers in Jane M. Oppenheimer, *Essays in the History of Embryology and Biology* (Cambridge, Mass.: The M.I.T. Press, 1967).

3. Chabry, "Contribution," pp. 306–09.

4. "Kürzer Bericht über die Sitzungen der vereinigten 5. and 9. Sektion für Zoologie und Anatomie der 60. Versammlung deutscher Naturforscher und Ärtze im Wiesbaden," *Anatomischer Anzeiger* 2 (1887): 763–4.

5. Wilhelm Roux, "Die Methoden zur Hervorbringung halber Froschembryonen und zum Nachweis der Beziehung der ersten Furchungsebenen des Froscheies zur Medianebene des Embryo," (1894), reprinted in Roux, *Gesammelte Abhandlungen über Entwickelungsmechanik der Organismen,* 2 vols. (Leipzig: Wilhelm Engelmann, 1895), cf. 2:957–9.

6. This would have been his "Beiträge zur Entwickelungsmechanik des Embryo. Nr. I. Zur Orientirung über einige Probleme der embryonalen Entwickelung," reprinted in *Ges. Abh.* 2:144–255.

7. See also Chabry, "Contribution," p. 237.

8. Ibid., p. 168.

9. Laurent Marie Chabry, "Note sur les monstres demi-individus latéraux," *Société de biologie, Comptes rendus* 38 (1886):323–5.

10. Laurent Marie Chabry, "Monstres nouveaux chez les ascidies," ibid., 37 (1885): 42–4.

11. See Edward Stewart Russell, *Form and Function, A Contribution to the History of Animal Morphology* (London: John Murray, 1916), p. 317.

12. Jane M. Oppenheimer, "Some Historical Relationships between Teratology and Experimental Embryology," *Bull. Hist. Med.* 42 (1968): 159.

13. Chabry, "Contribution," pp. 168–84.

14. There is an enormous literature in experimental teratology which in a very direct way can be traced back to the work of Etienne Geoffroy Saint-Hilaire in the first quarter of the century. I will have more to say about this tradition.

15. Chabry, "Contribution," pp. 300–01.

16. Ibid., p. 305.

17. Ibid., p. 309. Chabry did not keep figures on the number of experiments he attempted or on the number of eggs which he continued to develop after the operation. Casual remarks indicate that fewer than 30 lanced individuals were raised as far as gastrulation.

18. Roux, "Beiträge Nr. I. Zur Orientirung," *Ges. Abh.* 2:146–53.

19. Roux, "Half-Embryos," p. 9.

20. Ibid., pp. 12–23.

21. Chabry, "Contribution," p. 237.

22. Ibid., pp. 239–47. *Hémitérie* literally meant a simple monstrosity. Chabry explicitly denied using the classes as natural taxa.

23. Ibid., pp. 250–6.

24. Ibid., p. 250, emphasis is Chabry's.

25. Ibid., p. 253, emphasis is Chabry's.

26. Ibid., p. 256.

27. This orientation of Chabry's, that the experimental teratologist endeavors to copy nature, appears with peculiar emphasis at the end of this discussion.

Ce cas général [the distinction between normal and artificial monster] est important à bien saisir car il indique exactement la signification de chaque observation ou expérience qu'on peut faire sur les monstres et on n'est pas ainsi entraîné à exagérer l'importance de telle ou telle méthode tératogénique ce qu'ont fait quelques expérimentateurs qui, produisant des monstres par une intervention, sur des oeufs normaux, postérieure à la fecondation, ont cru en cela contrefaire en tous points la nature, alors qu'ils ne faisaient (comme moi-même) qu'imiter son procédé le plus simple. (Ibid., p. 256.)

28. Roux, "W. Roux," pp. 33–4.

29. In this discussion I will be following the outline given in Wilhelm Roux, "Die Entwickelungsmechanik der Organismen, eine Anatomische Wissenschaft der Zukunft" (1889), reprinted in *Gesammelte Abhandlungen* 2:24–54; and *Die Entwickelungsmechanik, ein Neuer Zweig der biologischen Wissenschaft* (*Vorträge und Aufsätze über Entwickelungsmechanik der Organismen*, ed., Wilhelm Roux, Heft 1 [Leipzig: Wilhelm Engelmann, 1905].)

30. Roux, "Entwickelungsmechanik der Organismen," p. 31.

31. Roux, *Neuer Zweig,* pp. 147–9.

32. *Terminologie der Entwickelungsmechanik der Tiere und Pflanzen,* ed. Wilhelm Roux et al. (Leipzig: Wilhelm Engelmann, 1912), p. 140.

33. Roux, *Neuer Zweig,* p. 15.

34. Chabry, "Contribution," pp. 261–2, 266.

35. Ibid., p. 312.

36. Ibid., pp. 256–60.

37. Ibid., p. 259.

38. Ibid., p. 260; see footnote, p. 238, for an explicit disclaimer on Chabry's part that he has created a natural system of classification.

39. See for example, Wilhelm Roux, "Für unser Programm und seine Verwirklichung," *Archiv für Entwickelungsmechanik der Organismen* 5 (1897):313–14.

40. Roux, "Zur Orientirung," in *Ges. Abh.* 2:154–5.

41. I wish to thank M. J. S. Rudwick for making this distinction explicit.

42. From English translation, Roux, "Half-Embryos," p. 4.

43. Chabry, "Contribution," pp. 298–302.

44. Ibid., p. 298.

45. Ibid., p. 299.

46. Ibid., pp. 210–13.

47. Ibid., p. 270.

48. He did, however, have a lively interest in the relationship between the first cleavage plane and the medial axis of two species of frogs. His work on this subject appeared in 1883, and it is curious that this was the only paper of Roux's which Chabry cited.

49. Roux, "Half-Embryos," p. 31.

50. Wilhelm Roux, *Ueber die Bedeutung der Kerntheilungsfiguren. Eine hypothetische Erörterung* (Leipzig, 1883), reprinted in *Ges. Abh.* 2:125–43.

51. Roux, "Half-Embryos," p. 27.

52. Ibid., p. 28.

53. Wilhelm Roux, "Beiträge zur Entwickelungsmechanik des Embryo. VII. Ueber Mosaikarbeit und neuere Entwickelungs-hypothesen," reprinted in *Ges. Abh.* 2:818–71.

54. Ibid., p. 821.

55. The reader will do best to read the two pertinent papers in order to verify for himself the emphasis I have given here: Wilhelm Roux, "Ueber die Verzweigung der Blutgefässe. Eine morphologische Studie" (1878), reprinted in *Ges. Abh.* 1:1–76; and "Ueber die Bedeutung der Ablenkung des Arterienstammes bie der Astabgabe" (1879), reprinted in *Ges. Abh.* 1:77–101.

56. Thomas S. Kuhn, *The Structure of Scientific Revolutions* (Chicago: University of Chicago Press, 1962).

57. Roux, "Entwickelungsmechanik der Organismen," p. 44. It is not clear that Roux had actually read Chabry by this time.

58. Wilhelm Roux, *Anatomischer Anzeiger* (Verhandlungen der Anatomischen Gesellschaft auf der sechsten Versammlung in Wien, vom 7–9 Juni 1892) 7 (1892): 39–40.

59. See "Postscript–1969," in Kuhn, *Structure of Scientific Revolutions,* 2nd ed. (Chicago: University of Chicago Press, 1970); and Thomas S. Kuhn, "Reflection on My Critics" in Imre Lakatos and Alan Musgrave, eds., *Criticism and the Growth of Knowledge* (Cambridge: Cambridge University Press, 1970). My paper was completed before reading Kuhn's more recent essays. In that Kuhn still envisions a psychological divide between his "disciplinary matrices," however, I find it hard to fit my analysis into a story about a "scientific revolution." Intellectual differences there certainly were, but I look in vain for a "Gestalt switch."

60. Jane M. Oppenheimer, "Historical Introduction to the Study of Teleostean Development," *Osiris* 2 (1936):124–48; and "Some Historical Relationships," *Bull. Hist. Med.* 42 (1968):145–59.

61. Camille Dareste, *Recherches sur la production artificielle des Monstruosités ou essais de tératogénie expérimentale* (1st ed. 1877) (Paris: C. Reinwald Cie, 1891), pp. 1–46; and Leo Gerlach, *Die Entstehungsweise der Doppelmissbildungen bei den höheren Wirbelthieren* (Stuttgart: Ferdinand Enke, 1882), p. 3.

62. Armand de Quatrefage, "Tératologie et tératogénie," *Journal des savants* (1887):217–29, 251–365, 430–44; for quotation see p. 223. De Quatrefage endeavored to show that Dareste's work, which was under review, was qualitatively different from Geoffroy's. Despite differences among those two teratologists, the impression left by the review is just the reverse of what de Quatrefage intended!

63. Isidore Geoffroy Saint-Hilaire, *Histoire générale et particulière des anomalies de l'organisation chez l'homme et les animaux,* 3 vols. (Paris, 1832–1837), *passim.* For a revealing example of these efforts see the "Tableau général et méthodique des monstruosités," vol. 2, opposite p. 179.

64. See especially, Dareste, *Recherche,* pp. 237–38.

65. Ibid., pp. 557–9.

66. Geoffroy, *Histoire générale,* vol. 3, pp. 498–508.

67. Dareste, *Recherche,* p. 24.

68. Ibid., pp. 563–4.

69. Geoffroy, *Histoire générale,* 3:452–70. These pages comprise a chapter entitled "De la réduction des lois tératologique aux lois générales de l'organisation."

70. Ibid., p. 453.

71. Ibid., pp. 466.

72. Dareste, *Recherche*, pp. 41–6; quotation appears on pp. 41–2.

73. See Charles Darwin, *Animals and Plants under Domestication*, 2nd. ed., rev. (London: John Murray, 1890) 2:257–79, for favorable reference to Dareste. Darwin was also familiar with Isidore Geoffroy's *Histoire générale*.

74. Ibid., p. 563.

75. Franz Keibel, "Gustav Albert Schwalbe," *Anatomischer Anzeiger* 49 (1916):210–21. This obituary includes a full bibliography of Schwalbe's publications.

76. Roux, "Wilhelm Roux," pp. 3–4.

77. Ibid., p. 37.

78. Ernst Haeckel, *Generelle Morphologie der Organismen. Allgemeine Grundzüge der organischen Formen-wissenschaft, mechanisch begründet durch die von Charles Darwin reformirte Descendenz-theorie*, 2 vols. (Berlin: Georg Reimer, 1866). The following thumbnail sketch of Haeckel's world view is drawn principally from vol. 1, pp. 63–166. The reader should survey the entire work to get the full flavor and breadth of Haeckel's persuasion.

79. Ibid., 2:451.

80. See particularly Wilhelm Roux, *Der Kampf der Theile im Organismus. Ein Beitrag zur Vervollständigung der mechanischen Zweckmässigkeitslehre* (Leipzig: Wilhelm Engelmann, 1881). Haeckel praised this work as presenting an extension of some of his own thoughts. Ernst Haeckel, *Natürliche Schöpfungs-Geschichte*, 2 vols. (Berlin: Georg Reimer, 1889), 1:253–8. When Roux reprinted the work in 1895, he disavowed some of the specific details, especially Haeckel's view of inheritance and development. Roux, *Ges. Abh.* 1:139. See also Frederick B. Churchill, "August Weismann and a Break from Tradition," *Journal Hist. Biol.* 1:91–112 for a discussion of Haeckel's views of inheritance and development and a similar rejection of them by Weismann.

81. ["Wilhelm Thierry Preyer"], *Leopoldina, Amtliches Organ der Kaiserlichen Leopoldino-Carolinischen Deutschen Akademie der Naturforscher* 33 (1897):116–17; "Thierry William Preyer [*sic*.]," *Nature* 56 (1897): 296.

82. William Preyer, *Specielle Physiologie des Embryo, Untersuchungen ueber die Lebenserscheinungen vor der Geburt* (Leipzig: Th. Grieben, 1885).

83. Ibid., p. 511.

84. Roux, "Wilhelm Roux," p. 3.

85. I.e., see *Ges. Ab.* 1:139.

86. Wilhelm Roux, "Professor Dr. Gustav Born," *Archiv für Ent-*

wickelungsmechanik 10 (1900):256–62; Walter Gebhardt, "Gustav Born," *Anatomischer Anzeiger* 18 (1900): 139–43.

87. Thomas Hunt Morgan, *The Development of the Frog's Egg, an Introduction to Experimental Embryology* (New York: Macmillan, 1897), has a convenient account of the many experiments of 1882–1884.

88. Geoffroy, *Histoire générale*, 1:27.

89. Ibid., p. ix.

90. Ibid., p. 26.

91. Henri de Lacaze Duthiers, "Direction des études zoologique," *Archives de zoologie expérimentale et générale* 1 (1872):1–64. William Coleman has translated the last thirty-six pages of this essay in his *The Interpretation of Animal Form* (New York and London: Johnson Reprint Corporation, 1967), pp. 132–78. Unfortunately, it is the earlier pages which express in explicit fashion the assumptions being drawn out here.

92. Claude Bernard, *Rapport sur les progrès et la marche de la physiologie générale en France* (Paris, 1867). I have not had access to this report, but since I am interested here only in a reaction against Bernard this neglect should not be critical to the argument. See also Victor Coste, Claude Bernard, Gabriel Auguste Daubrée and Michel Eugène Chevreul, "Note sur le rôle de l'observation et de l'expérimentation en physiologie," *Comptes rendus hebdomadaires des séances de l'académie des sciences* 66 (1868):1278–88 for another expression of this controversy.

93. Lacaze Duthiers, "Directions," pp. 7–13.

94. Ibid., p. 15. Von Baer, E. Geoffroy, Owen and Haeckel among others were associated with Cuvier's stage.

95. Ibid., p. 17.

96. Ibid., p. 21.

97. Ibid., p. 23.

98. Ibid., p. 24.

99. Ibid., pp. 39–40; see Coleman's translation, p. 144, for this account and the quotation.

100. Chevreul became an assistant attached to the Museum of Natural History in 1810 when Etienne Geoffroy was still in his prime! He rose to the chair of chemistry in 1830 and became a renowned authority on the chemistry of dyes. He was also interested in applications of chemistry to public health, therapeutics, and agriculture, among other things. He wrote two extensive works on scientific method: *Lettres adressées à M. Villemain . . . sur la méthode en général* (Paris: 1856); and *De la méthode à posteriori expérimentale* (Paris: 1870); see A. M. [Alfred Maury?], "M. Chevreul," *Journal des savants* (1899): 249–52. I will be following a shorter account of his method: M. E. Chevreul, "Considération sur la philosophie et application à la médecine d'une méthode

employée à rechercher la cause des différences que présentent les eaux naturelles dont on fait usage en teinture," *Journal de l'anatomie et de la physiologie* 1 (1864):1–26.

101. Ibid., p. 13.

102. "Chaque branche se divise en rameaux, dont chacun représente une méthode, et l'ensemble des rameaux et la branche à laquelle ils s'unissent représentent la philosophie de la science, que cette branche représente elle-même" (Ibid., p. 9).

103. Pouchet, "Laurent Chabry," *Journal de l'anatomie et de la physiologie* 29 (1893): 735–9.

104. Roux, "Ziele und Wege der Entwickelungsmechanik," *Ges. Abh.* 2:83.

105. Roux, "Wilhelm Roux," pp. 160–1.

106. Wilhelm Roux, "Für unser Programm und seine Verwirklichung," *Archiv für Entwickelungsmechanik der Organismen* 5 (1898): 313.

8 Statistics and Social Science

VICTOR L. HILTS
University of Wisconsin

In December 1888 the English student of heredity, Francis Galton, published in the *Proceedings of the Royal Society* a paper entitled "Co-relations and their Measurements, Chiefly from Anthropometric Data." In this paper Galton explained how it was possible to utilize a single number in order to measure the degree of co-relation of two phenomena, explaining what he meant in the following terms:

> Two variable organs are said to be co-related when the variation of the one is accompanied on the average by more or less variation of the other, and in the same direction. Thus the length of the arm is co-related with that of the leg, because a person with a long arm has usually a long leg, and conversely. If the co-relation be close then a person with a very long arm would usually have a very long leg; if it be moderately close then the length of his leg would be only long, not very long; and if there were no co-relation at all then the length of his leg would on the average be mediocre. It is easy to see that co-relation must be the consequence of the variations of the two organs being partly due to common causes. If they were wholly due to common causes, the co-relation would be perfect, as is approximately the case with the symmetrically disposed parts of the body. If they were in no respect due to common causes, the co-relation would be *nil*. Between these two extremes are an endless number of intermediate cases, and it will be shown how the closeness of co-relation in any particular case admits of being expressed by a simple number.[1]

It is fair to say that the idea which Galton expressed in his paper of 1888 and the developments which that idea stimulated created a revolution in the methodology of the social sciences. Within five years of the publication of his paper Galton had found a mathematical disciple in the person of the professor of applied mathematics at University College, London—Karl Pearson. During the last decade of the nineteenth century, Karl Pearson fashioned a new mathematical statistics which was eventually to change the character of almost all social science. If the social sciences of the mid-twentieth century have reached the point that no graduate student in economics, sociology, psychology, or even history can afford to neglect the importance of statistical methods, the credit (or if one wishes, the blame) is ultimately traceable to the influence of Galton and Pearson. In fact, I think it might even be fair to say that the introduction of statistical techniques in the social sciences represents one of the two most important methodological innovations of nineteenth-century science; the only nineteenth-century contribution to scientific method of comparable importance in terms of its impact upon many rather than a single discipline was that of the research laboratory. Just as the scientific laboratory spread from very modest beginnings to be a fixed feature of all modern science, so have statistical techniques made their mark almost everywhere.[2]

Why should Francis Galton have been the man to revolutionize the use of statistics in the social sciences? At first sight Galton might seem an unlikely candidate for such a role. Although he was an extremely bright and versatile scientist and a cousin of Charles Darwin, Galton was not a trained mathematician and his primary interests lay in the fields of anthropology, biology, and psychology. In fact, however, it was just this detachment from the traditional problems of contemporary statistics which proved fertile.

It is well known that Galton's most important predecessor in statistics was the famous Belgian statistician Adolphe Quetelet. Quetelet's lasting reputation is based upon his discovery that human stature is distributed in many populations according to the law of error. This was the first indication that the error distribution which had been used by Laplace and Gauss in connection with errors of astronomical observations had a much more general application. Several historians have already realized, however, that Quetelet never broke with the traditional conception that the law of error really is a law of error and not a more general law of distribution. Significantly this meant that for Quetelet there was no possibility of making any sort of further analysis of any distributions which obeyed the famous exponential formula. Galton, however, really did see the law of error as a more general law of distribution. In looking closely at Quetelet and Galton one can see how the deep differences which divided them in their understanding of the error law evolved from the particular problems with which they were involved. Whereas much of Quetelet's effort was spent in trying to make concrete the idea of the average, through the conception of the "average man" at the center of "social physics," Galton could not have been less interested in averages, certainly not the "average man." Galton originally became involved with statistics through his discovery that differences in ability and "genius" may be considered to be hereditary. Because Galton was interested in the inheritance of individual differences, he was led to focus his attention upon statistical deviations and variations as something important in their own right; and because of this, he was ultimately led to make his discovery of the correlation coefficient.

What I would like to do is to examine a crucial turn in the development of nineteenth-century statistical social science and to show how that turn was related to the development of the social sciences themselves. If the development of scientific method affects the development of science, the converse is also true.

QUETELET AND THE "AVERAGE MAN" CONCEPT

In a way the intellectual biography of Adolphe Quetelet is that of a novelist whose hero takes on an independence of his own and eventually spoils the plot.[3] The basic task of statistics during most of the nineteenth century was the use and interpretation of statistical averages. In order to provide a plot for the discussion of statistical averages, Quetelet in the eighteen-thirties created a new science which he called "social physics." The chief hero of this social physics was the "average man." What happened was that the hold of the average man concept on Quetelet became so great that the very discovery which might have attracted his attention away from averages and to the importance of deviations from the average—the discovery that the law of errors applies to a wide range of statistical regularities—was seen only as further evidence for the importance of the average. In the end, thus, the average man whom Quetelet introduced as the hero of his social physics became the villain who kept Quetelet from understanding the full implications of his own most important discovery.

In order to understand the hold which the average man concept held over Quetelet, it is necessary to examine in some detail the way in which the early development of statistics led naturally not only to an emphasis upon the average but also to the development of a philosophy which justified that emphasis.

If one tries to trace back the roots of Quetelet's conception of statistics, it is clear that one of the most important influences was the theorem proven early in the eighteenth century by the great Swiss mathematician Jacques Bernoulli. Although at the most obvious level it might be said that Bernoulli's theorem was simply a mathematical statement concerning probabilities and their frequencies, at another level it could be interpreted as a theory in defense of inductive reasoning or as a particular kind of epistemology. Mathematically the theorem stated, in very simplified language, that an event which occurs with a certain probability appears with a frequency approaching that probability as the number of observations is increased. Laplace expressed Bernoulli's theorem in the following terms in his *Essai philosophique sur les probabilités:* "in multiplying indefinitely the observations and experiences, the ratio of the events of different natures approaches that of their respective probabilities in the limits whose interval becomes more and more narrow in proportion as they are multiplied, and becomes less than any assignable quantity."[4]

Crucial to the way in which Bernoulli's theorem influenced Quetelet was the fact that Laplace used the theorem to give a high degree of concreteness to statistical regularities. According to Laplace there is an identity between the results of the constant repetition of observations and the *constant causes* acting in nature. Even in the field of history it should be possible, thought Laplace, to separate out the important constant causes:

> History treated from the point of view of the influence of constant causes would unite to the interest of curiosity that of offering to man most useful lessons. Sometimes we attribute the inevitable results of

these causes to the accidental circumstances which have produced their action. It is, for example, against the nature of things that one people should ever be governed by another when a vast sea or distance separates them. It may be affirmed that in the long run this constant cause, joining itself without ceasing to the variable causes which act in the same way and which the course of time develops, will end by finding them sufficiently strong to give to a subjugated people its natural independence or to unite it to a powerful state which may be contiguous.[5]

It had been the great achievement of Laplace as an astronomer to show that many of the particular phenomena of the solar system could be reduced to the constant cause by which the planets were affected, according to Newton's laws, by the gravitation of the sun, and the particular disturbing causes arising from the mutual influences of the planets, their satellites, and comets. Not unnaturally, just as Laplace was able to show the universality of Newton's laws, he seemed to feel that the moral world too would have its universally operative constant causes—causes which like many astronomical phenomena could only be studied by the numerous multiplication of observations.

An example of Laplace's ability to indicate the importance of what he called constant causes in the face of what would appear to be individual exceptions is given by his very penetrating analysis of the reasons that the ratio of births (actually baptisms) of boys and girls was different during the last part of the eighteenth century in Paris than in France in general.[6] Laplace first noted that the ratio of births of boys to girls was 22 to 21 in France as a whole but 25 to 24 in Paris during the period from 1745 to 1784. By assuming "that we may compare the births to the drawing of balls from an urn which contains an infinite number of white balls and black balls so mixed that at each draw the chances of drawing ought to be the same for each ball," Laplace calculated that "it is a bet of 238 to 1" that the difference in the birth ratios is not due to chance alone. This ratio, he said, is "sufficient to authorize the investigation" of the particular cause at work. "It has appeared to me that the difference observed holds to this, that the parents in the country and the provinces, finding some advantage in keeping the boys at home, have sent to the Hospital for Foundlings in Paris fewer of them relative to the number of girls according to the ratio of births of the two sexes." Laplace then confirmed his conjecture by noting that the number of boys exceeded the number of girls at the Foundling Hospital by only 1/38th, and that when the figures at the Hospital were neglected, the ratio of births of boys to girls at Paris was 22 to 21, the same as that for the rest of France. In proving that the ratio of births of boys to girls was actually the same in Paris as elsewhere, although the official figures were different, Laplace was effectively presenting evidence that unknown constant causes which resulted in this ratio were always present.

Whatever the philosophical merit of Laplace's appeal to the dominance of constant causes, it is obvious that his methods when suitably applied could be very powerful tools indeed; his determination by probabilities that "it is a bet of 238 to 1" that the difference in birth ratios in Paris and France as a whole is not due to chance is an excellent model of statistical reasoning. Unfortunately, however, it was not a model which was sufficiently emulated by most early nineteenth-century statisticians. Most statisticians of the early nineteenth century, being primarily practical men, would have proceeded immediately to speculate about the reasons for the difference between birth ratios at Paris and in France as a whole without bothering to check, as Laplace did, whether the difference was sufficient to "authorize the investigation." With the exception of Joseph Fourier and Quetelet it was not until at least the middle of the century that statisticians began to be concerned about the significance of observed differences, and even Quetelet did not often bother to make such a check. Generally speaking it was easier and more tempting to suggest possible constant causes for observed differences than it was to calculate whether the differences themselves were really important.

The Danish statistician and historian of statistics Harald Westergaard has referred to the two decades after the death of Laplace as the "era of enthusiasm" for statistics.[7] The statistical enthusiasm of this period was not to any great extent derived from the influence of Laplace, the possibility of applying the theory of mathematical probability to a new realm, or the belief that constant causes would be revealed by the repeated observation of events. For the most part the contributors to the statistical enthusiasm were driven to their interest in statistics, either in an official or an unofficial capacity, by their involvement with particular problems. Of especial historical significance as a link between Laplace and Quetelet were the group of statisticians centered in Paris. The most important member of this group was the student of public health Louis René Villermé, the editor and one of the principal contributors to the *Annales d'Hygiène*.[8] Villermé had been trained as a physician but early abandoned practice and turned his attention to the relationship between health and social class. Particularly important was Villermé's anthropometrical investigation into the relationship between class and human stature, from which he came to the conclusion that stature is greater "in proportion as the country is richer, the comfort more general, houses, clothes, and nouishment better, and privations during infancy and growth less."[9] Another important statistician of the Paris group was the lawyer André Michel Guerry, who is usually credited with having coined the term "moral statistics" as a companion term to vital statistics for designating such things as criminal and educational statistics. Guerry's most important work was entitled *Essai sur la statistique morale,* published in 1833. In this work Guerry noted, as Quetelet did almost simultaneously, that the regularity of criminal statistics "is not to be attributed to chance."[10]

The newly available social statistics filled at least partially an information gap which had been lacking in the earliest part of the nineteenth century and made it possible for someone acquainted with the theory of probability to carry out and develop what had already been suggested by Laplace. This role was assumed by Adolphe Quetelet. With an initial training in mathematics, Quetelet's immediate inspiration for the study of probability theory was a direct contact with the French mathematicians in Paris in 1823. Quetelet had gone to Paris at this time because he had been appointed director of the new Brussels Observatory and wanted to learn French astronomical methods. After returning to Brussels, Quetelet also became involved, at the request of the government, in purely statistical work. The result of these two influences was his theory of social physics.

The new collection of social statistics made evident certain regularities which could not have been apparent to Laplace. Just as the classification of births by sex brought out certain constancies in the ratio of births of males and females, so did other classifications bring out other regularities. As with most contemporary statistics these regularities were generally expressed in averages or rates, differentiated by location or time. Thus Villermé examined the average stature and the average life expectancy of persons from different areas of Paris, while Guerry paid attention to the number of crimes of different sorts committed in different districts of France by persons of different ages. The natural assumption was that differences in such averages were in some way connected to the differences in the respective groups of the population. Thus Villermé, and later the English sanitary reformer Edwin Chadwick, based the scientific evidence in favor of improved sanitation upon the difference in life expectancies of persons in the different areas of metropolitan Paris and London.[11]

Quetelet was particularly struck by the averages in criminal statistics. It might have seemed that anyone who had understood the full implications of Bernoulli's theorem would not have been surprised to discover statistical constancies in the statistics of crime as well as vital statistics. But there was definitely an extra poignancy about the regularity which appeared from the criminal tables. Here certainly one was leaving the realm of natural causation and entering the world of moral causation. At any rate, in spite of his familiarity with probability theory and its possible explanation of the facts, Quetelet was impressed with the pathos of situation:

> Sad condition of the human species! The toll of the prisons, of the chains and of the scaffold seems fixed for it with as much probability as the revenue of the state. We are able to enumerate in advance, how many individuals will stain their hands with the blood of their fellows, how many will be forgers, how many prisoners, nearly as one is able to enumerate beforehand the births and deaths which must take place.[12]

Having begun to think about statistics during the late 1820s, Quetelet published his first famous book in 1835, *Sur l'homme et le développement de ses facultés, ou essai de physique social.* It was in this work that Quetelet introduced his "average man"—"l'homme moyen." The concept of the average man gave Quetelet a way of dispensing with the need for considering particular individuals. "It is the social body, which forms the object of our researches, and not the peculiarities distinguishing the individuals composing it," wrote

Quetelet in the introduction to his work.[13] In terms of probability theory the creation of the average man was defended by an appeal to Bernoulli's theorem. *"The greater number of individuals observed, the more do individual peculiarities, whether physical or moral, become effaced, and leave in a prominent point of view the general facts, by virtue of which society exists and its importance is preserved,"* Quetelet wrote, underlining the sentence to be sure that its importance was understood.[14]

At one point in *Sur l'homme,* Quetelet characterized the average man as a "fictitious being."[15] Yet this fictitious being was given a high degree of reality by Quetelet and was characterized as the being "for whom everything proceeds conformably to the medium results obtained for society in general."[16] An analogy between the average man and the center of gravity gave Quetelet a way of considering the effect of changes in society upon society itself, those things which resulted in a change in the average being considered as "disturbing causes." Thus Quetelet wrote that "the purpose of this work is to study in their effects the causes, whether natural or disturbing, which influence human development; to endeavor to measure the influence of these causes, and the mode according to which they mutually modify one another."[17] Largely because of Quetelet's discussion of the effect of disturbing causes, one student of his ideas, Frank Hankins, has interpreted Quetelet as anticipating the statistics of correlation.[18] This is true only to a very limited degree, however. Practically speaking Quetelet's analysis meant no more than the kind of comparing and contrasting of averages in different populations which was typical of the practical statisticians. Because Quetelet looked upon the influence of disturbing causes in terms of their influence upon the center of gravity of the social system, he could not foresee Galton's concept of correlation, which involved the distributions as well as the averages.

It may be said that the net result of Quetelet's use of the average man "around which oscillate the social elements" was to give a much greater importance to the average than would have been required even by Bernoulli's theorem. And having once formulated such a grand interpretation of statistical averages, it was very difficult, in fact almost impossible, for Quetelet ever to back up and focus his attention upon the importance of the individual or of deviations from the average. Indeed, the remaining history of Quetelet's intellectual development led to still further concretization of the average man.

In spite of Quetelet's emphasis upon the average man in *Sur l'homme,* there has been some debate among students of Quetelet's writings about whether he was or was not interested in figures pertaining to the individual. Most early students of Quetelet assumed that he was not. More recently, however, the sociologist Paul Lazarsfeld has rejected this interpretation.[19] In support of his opinion that Quetelet was interested in individual measurements, Lazarsfeld has cited Quetelet's discussion in the third book of *Sur l'homme* on the measurement of the moral faculties. A close reading of the passages concerned, however, shows that Quetelet discussed the measurement of individuals in *Sur l'homme* only because individual measurements seemed necessary in order to find averages. It was easy to determine the average man with respect to physical characteristics like height, because one could use a ruler to measure stature, but it is not so obvious what to do in respect to "genius, prudence, or evil propensities." Realizing this difficulty, Quetelet expressed the hope that "in a more perfect state" courageous and virtuous actions would be recorded as routinely as crimes, and that one would then at least be able to "study the relative degrees of courage or virtue at different ages"—just as he himself had studied average numbers of crimes committed by persons of different ages.[20] Nowhere, however, did he indicate in *Sur l'homme* that the individual deviations themselves should be the thing of principal interest, and in fact he rejected the idea that one might ever find the proper *unit* for the measurement of moral qualities:

It appears to me that it will always be impossible to estimate the absolute degree of courage, &c. of any one particular individual: for what must be adopted as unity?—shall we be able to observe this individual long enough, and with sufficient closeness, to have a record of all his actions, whereby to estimate the value of the courageous ones; and will these actions be numerous enough to deduce any satisfactory conclusion from them? Who will guarantee that the dispositions of this individual may not be altered during the course of the observations? When we operate on a great number of individuals, these difficulties almost entirely disappear, especially if we only want to determine the ratios, and not the absolute values.[21]

It might have been thought that Quetelet's discovery that anthropometrical measurements are distributed in the same way as errors in astronomy might have led him to change the focus of his attention from the average to the deviation from the average and to recognize the law of error as a law of distribution. In fact, however, nothing of the sort occurred. What Quetelet did was to superimpose his discovery concerning the distribution of anthropometrical measurements upon his theory of the average man in such a way as to reinforce his earlier theory.

In order to understand Quetelet's interpretation of the law of error one must examine his analysis of causation as presented in his *Lettres sur les probabilités*. It was this work, first published in 1845, which contained Quetelet's most systematic discussion of the various applications of the law of error, and which ultimately became influential upon Francis Galton. The entire third part of the book, in which Quetelet introduced the law of error, is entitled "The Study of Causes." Here Quetelet distinguished between three kinds of causes: constant causes, variable causes, and accidental causes. The definitions which Quetelet gave for the three causes were the following:

Constant causes are those which act in a continuous manner, with the same intensity, and in the same direction.
Variable causes act in a continuous manner, with energies and tendencies which change either according to determined laws or without any apparent law. Among variable causes, it is above all important to distinguish such as are of *periodic* character, as for instance the seasons.
Accidental causes only manifest themselves fortuitously and act indifferently in any direction.[22]

The important distinction is between the accidental causes and the other two kinds of causes, since variable causes are essentially like constant causes except for their time dependency.

What Quetelet showed in his *Lettres sur les probabilités* is that, under suitable assumptions, the effects of accidental causes distribute themselves according to the law of error around a mean which is itself in some way determined by the constant (or variable) causes. The specific model which Quetelet had in mind, and from which he developed his mathematical proof that the effects of accidental causes would distribute themselves according to the law of error, was that of drawing black and white balls from an urn. In this model it is the accidental causes which determine which balls are drawn on a particular occasion and the constant causes which determine the ultimate ratio of black and white balls after a large number of drawings.[23] There are two important implications of this. First, the law of error remains essentially a law of error because it is developed by the accidental causes. Second, there is no possibility of discovering anything about the important constant causes in nature from the character of the error distribution, since this distribution is related only to accidental causes: in order to understand the action of constant causes one continues to look at the mean. Although the error distribution tells us something about the precision of our knowledge, it does nothing more.

Because of Quetelet's interpretation of the law of error an entirely new distinction was made in *Lettres sur les probabilités* between ordinary arithmetical means, which could be formed of any group of numbers, and true or "typical" means, which refer to phenomena whose distributions conform to the law of error. Using the word "average" instead of arithmetic mean, the distinction was perhaps better described in a review of the *Lettres sur les probabilités* in 1850 by the English astronomer John Herschel than it was by Quetelet himself:

An average may exist of the most different objects, as the heights of houses in a town, or the sizes of books in a library. It may be convenient to convey a general notion of the things averaged; but it involves no conception of a natural and recognizable central magnitude, all differences from which ought to be regarded as deviations from a standard. The notion of a mean, on the other hand, does imply such a conception, standing distinguished from an average by this very feature, viz. the regular marching of the groups, increasing to a maximum and thence again diminishing. An average gives us no assurance that the future will be like the past. A mean may be reckoned on with the most implicit confidence.[24]

In effect this was simply another way of stating the importance of an average around which there is an error distribution and, therefore, another way of indicating the reality of the mean in such cases. In the words of the writer of an Austrian book on statistics which was used in translation as a textbook in the United States as late as the 1910s: "The typical means are independent scientific perceptions. The series of items compared with the typical means thus loses the greatest part of its importance. It is worthy of consideration only as a measurement of variability."[25] In terms more in keeping with Quetelet's terminology, deviations from typical means can be due only to accidental causes.

Quetelet introduced his discovery that the heights of men in some populations are distributed in conformity to the law of error by first considering all the errors which might be made when taking measurements of a statue of a gladiator. Of necessity such measurements would follow the law of error. It might be objected, Quetelet then interjected, that no one would make a large number of such measurements. Yet, and here came the revelation, the measurements had already been made. "Yet, surely, more than a thousand copies have been measured, which I do not assert to be that of the Gladiator," said Quetelet, "but which in all cases differs but little from it." The measurements to which he referred were those reported in the thirteenth volume of the *Edinburgh Medical Journal* on 5738 soldiers in the Scotch regiments.[26] Quetelet showed that these measurements followed very closely what should have been predicted had the measurements in fact been made upon one single statue with measuring errors distributed according to the law of error.

The conclusion which Quetelet drew from his data on the Scotch soldiers was in complete harmony with his understanding of the law of error, the distinction between constant and accidental causes, and the entire process by which a gradual concretization of statistical averages had occurred since Laplace's introduction of the concept of constant cause. In *Sur l'homme,* Quetelet had assumed that the average man had a great deal of reality, but now he thought that he had proof of that reality, because the average man is in fact simply a representative of the human *type* itself:

Of the admirable laws which Nature attaches to the preservation of the species, I think I may put in the first rank that of maintaining the type. . . . I have already endeavoured to determine this type of knowledge of the human mean. But if I mistake not, what experiment and reasoning had shown men, here takes the character of a mathematical truth.

The human type, for men of the same race, and of the same age, is so well established that the differences between the results of observation and of calculation, notwithstanding the numerous accidental causes which might induce or exaggerate them, scarcely exceed those which unskillfulness may produce in a series of measurements taken on one individual.[27]

All of this had one extremely important implication; it meant that there could be no science of the study of individual deviations from the average. Quetelet's discovery that the average man represents the human type was the final stage in his concretization of the statistical average. The average man represents a true mean and the differences from this mean can only be the result of accidental causes—they are, in other words, fundamentally unanalyzable.

With Quetelet's discussion of the application of the law of error to human stature one can see both his contact and his contrast with Francis Galton. Whereas Quetelet saw the law of error as law of error and nothing more, Galton would see it as simply a law of distribution. Whereas Quetelet implied that there could be no science of the deviation of man from the mean—on the ground that such deviations were the result of accidental causes—Galton was to show that there was such a science, and that it was concerned with the laws of heredity. Whereas Quetelet talked about the effect of constant causes upon averages in general, and to this extent was dealing with correlations, Galton was to show that it made sense to speak about correlations as affecting the distribution itself.

FRANCIS GALTON AND THE STATISTICS OF "INDIVIDUAL DIFFERENCES"

During the middle of the nineteenth century a great shift in emphasis occurred during which biological and evolutionary ideas came to dominate much of social science. In spite of his interest in anthropometry, however, Quetelet never modified his social physics to take into account this new emphasis. His social physics had crystallized during the 1830s at the height of the "era of enthusiasm" for statistics, when faith was placed upon moral and social, not biological, development. Significantly, this meant an almost complete omission in Quetelet's writings of any discussion of heredity as one of the important constant causes. Had Quetelet considered heredity one of the important factors, it might not have been so easy for him to consider anthropometrical variation as simply the result of accidental causes, but for him the contradiction did not arise. Here is where Francis Galton took over.

It would be interesting to discuss in some detail the way in which Francis Galton first became convinced of the importance of heredity in determining human ability and character, but there is no room to do this here.[28] Galton's first statement of his hereditarian thesis occurred in a short paper published in 1865 in *Macmillan's Magazine*.[29] Because this was six years after the publication of Charles Darwin's *Origin of Species,* there is always a temptation to interpret Galton's ideas as a direct response to those of Darwin. But this is too simple; Galton said he was encouraged to pursue his own ideas by the publication of the *Origin of Species,* but he also indicated that these ideas had "long interested" him.[30] Something, but not much, about the other influences upon Galton is known. In the early 1860s Galton was engaged in an ethnological study of the mental peculiarities of different races, and in this connection was thinking about things which could influence national and racial character—one such thing being heredity. Probably before the publication of the *Origin of Species,* Galton had already become intrigued with the idea that the human race could be improved by the regulation of marriage, an idea which involved heredity and which had long historical roots. Finally, Galton encountered writings on hereditary diseases and other subjects, such as that given by the English author G. H. Lewes on the basis of a French treatise on the subject by Prosper Lucas.[31] Such a brief account of the influences at work on Galton's mind just before 1865 does not do justice to the true complexity of the situation, however.

Besides the just mentioned scientific influences upon Galton during this time, it is important to take into consideration Galton's great feeling as a traveler for the reality of human differences. As a youth he had visited the Near East and, as he said, "revelled in the glory" of Constantinople, where he heard stories "about a phase of humanity which I did not esteem but was glad to know about."[32] In 1860 Galton visited Spain with a scientific expedition, and was delighted with the uniqueness of the Spanish people:

> It was a great delight to me to find that the Spanish ways of life appeared thoroughly characteristic, and wholly uncopied from other nations of modern Europe. There is a common cant phrase used sometimes in respect to France, and sometimes to England, of "advancing in the van of European civilization." Yet, however flattering to our vanities, it would be a matter of deep regret if European civilization should ever become so far one and indivisible, that nations, whose instincts and geographical conditions of life are different, should make it a point of fashion or of education to live on the same model. One longs to see a freer development than exists at present, of the immense variety of aptitudes and peculiarities that are found in the human race, and are fostered by different geographical circumstances.[33]

Beside this enthusiastic description of Spanish national character, one may juxtapose a quotation written in a similar spirit from Galton's paper of 1865 on "Hereditary Talent and Character."

> How enormous is the compass of the scale of human character, which reaches from dispositions like those [of the murderer Townley], to that of a Socrates! How various are the intermediate types of character that commonly fall under everybody's notice, and how differently are the principles of virtue measured out to different natures! We can clearly observe the extreme diversity of character in children. Some are naturally generous and tricky; some are warm and loving, others cold and heartless; some are meek and patient, others obstinate and self-asserting; some few have the temper of angels, and at least as many have the tempers of devils.[34]

No matter what his subsequent devotion to statistics, it can fairly be said that there was no danger of Francis Galton's falling under the spell of the average man. It is, of course, true that Quetelet also had recognized the existence of diversity among individuals, but not with such depth of feeling. For Quetelet individual differences were embarrassments to be gotten rid of by focusing upon averages and large numbers; for Galton they were almost the only thing of interest.

Together Galton's hereditarian thesis and his very great sensitivity to human differences was to lead him into new territory in many fields. Statistics first became involved in his scientific wanderings when he attempted to prove the hereditarian thesis. Although Galton was to contribute a few papers simply on statistical points, he was never to be much interested in statistical methodology for its own sake, but was always drawn to statistics by what it could contribute to his other interests. His first statistical evidence in favor of his hereditarian thesis was characteristically as crude as the information at hand. In 1865 Galton relied exclusively upon an analysis of biographical dictionaries in order to prove the importance of heredity as a cause of individual differences. In effect, all that he did was to compare the number of relatives mentioned in certain biographical dictionaries with the number of relatives which one might have expected on a chance basis. Four years later, in 1869, Galton published his famous *Hereditary Genius,* in which the same kind of evidence was used, although much more systematically. But in fact, as Galton himself realized, he was not able really to prove the hereditarian thesis by this means; all that he actually could prove was the existence of a surprising number of very gifted families. The hereditarian thesis followed only if one were also willing to assume that the existence of these gifted families was due to heredity and not to social factors. Realizing the weakness of his statistical arguments in favor of his thesis, Galton turned to biology, where he was more successful, and eventually found his best evidence in the contrast between identical and fraternal twins.

While Galton was not able to use statistics to give a convincing demonstration of the importance of heredity as the cause of individual differences in *Hereditary Genius,* he was able to use statistics there in an entirely new way in order to measure such differences. It was with this insight that he began his contributions to statistical methodology. In measuring differences in ability he was tackling the problem which Quetelet had already renounced in *Sur l'homme.* It will be remembered that Quetelet said "it appears to me that it will always be impossible to estimate the absolute degree of courage, &c. of any one particular individual: for what must be adopted as unity?" In England the same sentiment had been echoed by William Farr, the most knowledgeable student of vital statistics in England in the middle of the nineteenth century. "What are we to say to the human unit?" Farr had asked in an address to the statistics section of the British Association in 1864. Farr could give no really good answer but could only rephrase the question itself: "Nations differ in their intellect as well as their moral faculties; and the expression of these forces of the soul, whether we look at scientific achievements or vulgar errors, at virtues or crimes, is one of the difficult problems of statistics."[35] Galton was to show that the problem could be approached by the use of Quetelet's own law, the law of error.

In his autobiographical *Memories,* Galton tells us that his own introduction to the law of error came from his friendship with the mathematician, geographer, printer, and future president of the Royal Society, William Spottiswoode. Spottiswoode had published a paper in which he had utilized the law of error very much in the spirit of Quetelet although the subject matter was different. The title of Spottiswoode's paper, published in the *Journal of the Royal Geographical Society* in 1860, was "On Typical Mountain Ranges: an Application of the Calculus of Probabilities to Physical Geography." What Spottiswoode was trying to do was to provide a method by which it could be determined whether an entire mountain chain had been formed by one single cause. Spottiswoode reasoned that if the directions of the individual segments of the mountain chain could be considered to deviate around the general direction of the whole chain in accord with the law of error, it might be assumed that the whole chain arose from one cause. In his analysis Spottiswoode used the "scale of precision" included in Quetelet's *Lettres sur les probabilités.* To his friend Galton, Spottiswoode explained, as Galton recalled, "the far reaching application of that beautiful law."[36]

Galton said that he "fully apprehended" what Spottiswoode had to say.[37] In fact, he may have understood more than Spottiswoode told him. Whereas Spottiswoode applied the error law in traditional terms, according to which the action of constant causes is associated with the mean and an error distribution around the mean results only from accidental causes, Galton already had an entirely different idea in mind. Because Galton was able to associate the error distribution with individual differences caused by heredity, the distinction between constant and accidental causes lost much of its meaning. It is dubious, of course, that Galton himself immediately understood the entire significance of his break with tradition, and indeed, the end result of Galton's different veiwpoint only developed gradually.

In *Hereditary Genius,* Galton simply used the law of error in order to provide a scale of ability.[38] Galton did not actually prove that either ability or "genius" is distributed according to the formula $f(x) = [1/(\sigma\sqrt{2\pi})]\exp - \frac{1}{2}(x - a/\sigma)^2$. In fact the evidence which Galton introduced to justify his position was almost appallingly meager, the only empirical evidence being a series of examination scores from the mathematical tripos exam at Cambridge. In lieu of evidence Galton had to rely upon analogy: since the law of error applied to stature, it should apply to ability as well—a weak argument biologically and an incorrect one mathematically. The real reason that Galton introduced the law of error in *Hereditary Genius,* however, had nothing to do with either evidence or analogy: the real reason was simply that it gave a plausible way of transforming relative and comparative statements into absolute terms, and thus made it possible to offer a scale or measure of ability in the traditional sense. Galton's raw data stated things like "X men out of a population of Y achieve sufficient eminence to be listed in a biographical dictionary." What Galton wanted was a statement which would say something about the amount of ability, expressed in suitable units, possessed by the X men listed in the dictionary. By assuming that ability is distributed according to the law of error, Galton was able to pass from statements like "Bach had the musical ability of one in many millions" to statements like "persons possessing ability in excess of four times the probable error from the average are as uncommon as one in several million." Mathematically speaking, of course, Galton was simply passing from the cumulative distribution function of the error law to the law of error itself. But the important point was that Galton had found the unit in which ability could be measured—what previous writers had called the "probable error" and what Galton much preferred to call the "probable deviation."

If Galton had terminated his discussion of statistics with what appeared in *Hereditary Genius,* he would still have made a substantial contribution to the history of nineteenth-century statistics. It is true that Adolphe Quetelet indicated in some of his later writings that he, too, believed that intellectual as well as anthropometrical qualities would distribute themselves according to the error law, but Quetelet never developed the idea.[39] Furthermore, even if Quetelet had developed the idea, it would have been within terms of his own restricted understanding of the error law. Galton's difference from Quetelet became yet more manifest, however, in Galton's statistical writings after *Hereditary Genius.*

What carried Galton from *Hereditary Genius* to the discovery of the co-relation concept in 1888 was his study of the laws of heredity in terms of a model based upon the error distribution. In *Hereditary Genius,* Galton had shown how the assumption that ability follows the error law in its distribution allows one to set up a scale for the measurement of ability, the unit of measurement being the "standard deviation," or as Galton also referred to it, the "statistical unit." In some papers published in the 1870s, Galton generalized upon this by noting that one may set up a similar "statistical scale" for any particular quality which one assumes to be distributed according to the error law—at least in theory, if not in actual practice.[40] Assuming that every inheritable quality is, in fact, distributed according to the error law, it is thus possible to set up a scale for every inheritable quality. In all such scales, the degree to which a particular inherited quality is possessed by a particular individual is expressed in terms of the "statistical unit," which thus becomes the measurer of biological variation. Finally then—and it was this last point which Galton conceived to be one of his most fundamental insights—the laws of heredity must be concerned with deviations measured in statistical units. When Galton wrote his *Memories* in 1908, he said that he still clearly recalled the circumstances in which this last insight had been realized:

As these lines are being written, the circumstances under which I first clearly grasped the important generalisation that the laws of heredity are solely concerned with deviations expressed in statistical units, are vividly recalled to memory. It was in the grounds of Naworth Castle, when an invitation had been given to ramble freely. A temporary storm drove me to seek refuge in a reddish recess in the rocks by the side of the pathway. There the idea flashed across me, and I forgot everything else for a moment in my great delight.[41]

With Galton's insight in Naworth Castle the divergence between his understanding of the law of error and that of Adolphe Quetelet was complete. The ability of every man who is not of average ability represents a deviation from the average which may be expressed in statistical units, and similarly for all other inheritable qualities. Because parents of high ability usually have children of above average ability, the cause of any particular individual's being above average in ability is ascribable not solely to accidental factors but at least in part to heredity. Here, however, another insight occurred which led Galton even closer to the discovery of correlation. How large is that part which is ascribable to heredity? While it is true that parents of high ability, for example, will have children of above average ability, it is not necessarily true that any particular child will inherit all of his parents' ability. Nor is it even necessary that a particularly well-endowed couple will have children whose average endowment equals their own. What fraction, then, of the parental deviations is passed on to the offspring?

In 1877 Galton was able to give a theoretical solution to the question just asked by first assuming two things: 1) that the error distribution which describes the distribution of any particular quality in the population as a whole will be repeated from generation to generation; and 2) that if one considers the distribution of any quality among all children whose parents are equally endowed—a suitable average being taken of male and female contributions—this distribution too will be found to be an error distribution.[42] Galton introduced the term "mid-parent" to refer to the suitable average of mother and father, and he introduced the term "family" to refer to all children whose parents are similarly endowed. One more assumption was then made: that the variability of the error distribution representing each "family" is identical, regardless of the endowment of the parents. Effectively one thus has a statistical model of the population, in which two factors are at work. One of these factors is heredity and the other factor represents the combined influence of all non-inheritable causes. Because of heredity, a certain fraction of the parental endowment is passed on to the offspring, a fraction which Galton represented by "r" since it indicated the degree to which there is "reversion" to the mean. If reversion alone were the only factor, one would continue to have an error distribution from generation to generation, but its variability would become smaller and smaller. If one denotes the variability of the original population by "c," then the variability of the population after reversion alone would be rc. However, this centripetal tendency is counteracted by the centrifugal tendency which causes variability in each "family." Together the existence of reversion and the existence of family variability cancel one another out to the extent that the original population distribution is exactly maintained. Mathematically speaking, then, one has the maintenance of the original error distribution as the sum of two other error distributions, one of which has the variability rc and the other, the family, which has a variability which one may denote by v. Since the variability of the sum of two independent error distributions is the square root of the squares of the variabilities of the two distributions, one has the equation: $c^2 = (rc)^2 + v^2$.

Conceptually I think it is safe to say that by 1877 Galton had taken the major steps towards his discovery of the correlation coefficient through his analysis of the total variability in any population into two parts, that ascribable to heredity and that not so ascribable. Indeed, it might even be arguable that in 1877 Galton was beginning to penetrate into problems which twentieth-century statisticians would include under the analysis of variance. In 1877 Galton had shown that it was very meaningful to say that a certain fraction of the deviation or variability in a distribution represented by the law of error is "caused" by some one factor—although he had not shown that the procedure was more general than the particular hereditarian problems with which he was at the time concerned. The study of heredity was very directional in its character; children inherit from their parents, and not vice versa. But correlation is not directional and does not specify which is cause and which effect. Conceptually this was the only hurdle which Galton had yet to pass over.

During the 1880s Galton was engaged in comparing the stature of parents and offspring by preparing a two-dimensional table, with the heights of parents on one edge and the heights of offspring on the other edge. The table was blocked off into squares, and in each square Galton entered the number of individuals whose height was indicated by one coordinate and who had parents of a height indicated by the other coordinate. In fact the table represented a joint frequency distribution, as Galton compared the "number of entries in each square inch." Galton found it hard at first "to catch the full significance of the entries." Soon, however, a striking thing began to emerge: upon close inspection those squares in which there were the same number of individuals were seen to form ellipses. As before at Naworth Castle, again we have a moment of discovery:

> I had given much time and thought to Tables of Correlations, to display the frequency of cases in which the various deviations say in stature, of an adult person, measured along the top [of the table], were associated with the various deviations of stature of his mid-parent, measured along the side [of the table]. . . . But I could not see my way to express the results of the complete table in a single formula. At length, one morning, while waiting at a roadside station near Ramsgate for a train, and poring over a small diagram in my notebook, it struck me that the lines of equal frequency ran in concentric ellipses. The cases were too few for certainty, but my eye, being accustomed to such things, satisfied me that I was approaching the solution. More careful drawing strongly corroborated the first impression.[43]

What Galton had discovered was that his data had the characteristic elliptical property of the two-dimensional error distribution. It would be interesting to know if Galton felt any analogy between himself and Johannes Kepler some two and a half centuries earlier.

The discovery of the ellipses of equal frequency showed Galton that his coefficient of "reversion" also had another geometrical meaning. In poring over his tables, Galton noted that the children seemed to inherit about ⅔ of their parents' deviation from the average. But looking at the same table in another way, it might equally be said that the parents share ⅓ of their children's deviations. Furthermore, if the deviations for each are measured in terms of the statistical units, i.e., probable errors, then the two fractions will always be identical. In this case "r is a measure of the closeness of correlation." In 1889 in his book-length account of his researches on hereditary, *Natural Inheritance*—the book in which, incidentally, Galton finally decided to refer to the law of error as the "normal law"—Galton drew the ellipses and discussed the two fractions, but did not note the existence of a single correlation coefficient itself. Because he had been comparing the stature of the "mid-parent" with that of the offspring, and had been expressing each in terms of inches rather than statistical units for the purpose, the existence of a single number measuring correlation was not immediately obvious. However, when he converted to the use of statistical units, as he had been accustomed to doing since the 1870s, the correlation coefficient manifested itself. This is what he did in his publication in December 1888 of the paper on "Co-relations and their measurements, chiefly from Anthropometric Data" in the *Proceedings of the Royal Society*.

CONCLUSION

In conclusion it seems to me that it is possible to understand why Francis Galton and not Adolphe Quetelet should have been led to the idea of statistical correlation. In spite of the fact that Adolphe Quetelet was the first to show that the error law could be used to describe things other than errors, he was never able to liberate himself from the domination of the "average man" concept which had been at the basis of his social physics. Quetelet was very much a man of his own period, interested in the meaning of statistical regularities (mainly averages), the difference between constant and accidental causes, and the effect of moral and social change upon the human race. Quetelet did talk about the effect of changes upon society, and to this extent may have implied an idea of correlation, but it was a correlation concept that applied simply to averages and not to deviations.

As opposed to Quetelet, Francis Galton had no interest in the average man but great interest in men who are different from the average. Because of this difference in viewpoint, Galton was able to liberate himself from the opinion which interpreted deviations from the average in normal distributions solely as the result of accidental errors. Precisely because Galton did not approach statistics primarily from the mathematical viewpoint, which had incorporated the metaphysics of constant causes, he was able to develop statistical theory beyond the stage in which it had been left by Adolphe Quetelet. Galton himself was well aware that his approach to statistics was different from that of the mathematicians. This awareness was made especially evident in his discussion of the difficulty he had in getting technical help from mathematicians:

> He [a mathematician] helped me greatly in my first struggles with certain applications of the Gaussian Law, which for some reason that I could never clearly perceive, seemed for a long time to be comprehended with difficulty by mathematicians, including himself. . . . They were unnecessarily alarmed lest the well-known rules of Inverse Probability should be unconsciously violated, which they never were. I could give a striking case of this, but abstain because it would seem depreciatory of a man whose mathematical powers and ability were far in excess of my own. Still, he was quite wrong. The primary objects of the Gaussian Law of Error were exactly opposed, in one sense, to those to which I applied them. They were to get rid of, or to provide a just allowance for errors. But these errors or deviations were the very things I wanted to preserve and to know about. This was the reason that one eminent living mathematician gave me.[44]

The development of statistical methods in the nineteenth century shows the great difficulty of extending the mathematical methods of the physical sciences to the social and biological sciences without taking into account the particular problems of those sciences. Quetelet's social physics was an attempt to use the mathematical methods developed by astronomers as well as the scientific framework of physics and astronomy to create a new science of man. But Galton was more successful than Quetelet in developing mathematical methods appropriate to the social sciences because of his deeper insight into the science of human behavior itself. By approaching statistics with an awareness of individual differences and through the problem of hereditary variation, Galton was led into those studies which eventually resulted in his discovery of the correlation coefficient and thus to the beginning of modern mathematical statistics as used in the social sciences.

NOTES

1. Francis Galton, "Co-relations and their Measurements, Chiefly from Anthropometric Data," *Proceed. Roy. Soc.* 45 (1888):135–6.
2. For general history of statistics in the nineteenth century see

Helen M. Walker, *Studies in the History of Statistical Method, with Special Reference to Certain Education Problems* (Baltimore, 1929); Harald Westergaard, *Contributions to the History of Statistics* (London, 1932). Also see the author's 1967 Harvard Ph.D. dissertation, "Statist and Statistician: Three Studies in the History of Nineteenth-Century English Statistical Thought."

3. For Adolphe Quetelet, see Frank H. Hankins, "Adolphe Quetelet as a Statistician," *Studies in History, Economics and Public Law,* 21:4 (New York, 1908).

4. Laplace, *A Philosophical Essay on Probabilities,* trans. 6th French ed. F. W. Truscott and F. L. Emory (New York, 1902; reprinted, 1962), p. 187.

5. Ibid., pp. 63–4.

6. Ibid., pp. 67–9.

7. Westergaard, *History of Statistics,* ch. XIII.

8. For Villermé's connections with Quetelet, see Erwin H. Ackernecht, "Villermé and Quetelet," *Bull. Hist. Med.* 26 (1952): 317–29. For the general context, see George Rosen, "Problems in the Application of Statistical Analysis to Questions of Health: 1700–1800," *Bull. Hist. Med.* 29 (1955): 27–45.

9. Louis Villermé in *Annales d'Hygiène,* quoted by Adolphe Quetelet, *A Treatise on Man and the Development of his Faculties,* trans. R. Knox (Edinburgh, 1842), p. 59.

10. André Michel Guerry, *Essai sur la statistique morale de la France, Précédé d'un rapport a l'académie des sciences, par MM Lacroix, Silvestre et Girard* (Paris, 1833), p. vii.

11. The best study of Chadwick from the point of view of the history of statistics is M. W. Flinn, "Introduction," in Edwin Chadwick, *Report on the Sanitary Condition of the Labouring Population of Great Britain* (Edinburgh, 1965).

12. Quoted in N. F. Mailly, *Essai sur la vie et les ouvrages de L.-A.-J. Quetelet* (Brussels, 1875), p. 97.

13. Quetelet, *A Treatise on Man,* p. 7.

14. Ibid., p. 6.

15. Ibid., p. 8.

16. Ibid.

17. Ibid., p. 8.

18. Hankins, "Adolphe Quetelet as a Statistician." On p. 84, Hankins says of Quetelet that his moral statistics "attempts to correlate the phenomena under investigation with certain physical and social conditions, by showing variations in the numbers as the conditions are changed."

19. Paul F. Lazarsfeld, "Notes on the History of Quantification in Sociology—Trends, Sources and Problems," *Isis* 52 (1961): 172–9.

20. Quetelet, *Treatise on Man*, p. 73.

21. Ibid., p. 73.

22. Adolphe Quetelet, *Letters addressed to H.R.H. the Grand Duke of Saxe Coburg and Gotha, on the Theory of Probabilities, as applied to the Moral and Political Sciences*, trans. Olinthus G. Downes (London, 1849), pp. 107–8.

23. Ibid., p. 108.

24. John Herschel, "Quetelet on Probabilities," *Edinburgh Review* (July 1850), p. 23.

25. Frank Žižek, *Statistical Averages, A Methodological Study*, trans. Warren M. Persons (New York, 1913), p. 168.

26. Quetelet, *Letters on Probability*, p. 92.

27. Ibid., p. 93.

28. The major source of information on Francis Galton is the three volume biography by Karl Pearson, *The Life, Letters, and Labours of Francis Galton* (London, 1914–30).

29. Francis Galton, "Hereditary Talent and Character," *Macmillan's Magazine* (June 1865).

30. Francis Galton, *Memories of My Life* (London, 1908), p. 288. The following discussion of the genesis of Galton's hereditarian thesis follows in a general way the author's treatment in his dissertation, "Statist and Statistician."

31. George Henry Lewes, *Physiology of Common Life* (London, 1859).

32. Galton, *Memories*, p. 52.

33. Francis Galton, *Narrative of an Explorer in Tropical South Africa*, 4th ed. (London, 1891), pp. 220–1.

34. Galton, "Hereditary Talent and Character," p. 325.

35. William Farr, "Economic Science and Statistics," *BAAS Transactions* (1864), p. 158.

36. Galton, *Memories*, p. 304.

37. Ibid.

38. Francis Galton, *Hereditary Genius: An Inquiry into its Laws and Consequences* (1869) (New York, 1962), pp. 57–76.

39. See discussion in Hankins, "Adolphe Quetelet as a Statistician," pp. 75–6, 94.

40. In particular, Francis Galton, "Statistics by Intercomparison," *Philosophical Magazine* (Jan. 1875). Also see discussion in Pearson, *Life of Francis Galton*, 2:392–3. Pearson includes a picture of Galton's comparison of seed sizes on the basis of the statistical scale. In this Galton writes: "The gradations are those of the so-called 'statistical scale' based on the law of frequency of error. It is as follows:—when the seeds are set in a row, *in order of their sizes*, those at or near the middle of the row are ranked as of $0°$ (these are of average size); those at or

near the quarter points as of $\pm 1°$; those at or near one twelfth from either end, as of $\pm 2°$."

41. Galton, *Memories,* p. 300. Karl Pearson seems to have misunderstood this, and thinks that it was related to a discovery which Galton made in 1888 or 1889 just before his final discovery of the correlation coefficient. From Galton's discussion in the *Memories,* however, it is clear that the discovery really belongs at this earlier period, and only in so interpreting it can one make sense of Galton's development. Pearson also contradicts himself in two volumes of his biography. In vol. IIIa: 50, he assigns the date to 1888, just before the discovery of the correlation coefficient; in vol. II: 393, he assigns it to 1889, which would be after the discovery of the correlation coefficient.

42. Francis Galton, "Typical Laws of Heredity," *Proc. Roy. Inst.* 8 (February 11, 1877). Although not related to the subject of the present paper it is interesting to note that in one sense Galton did accept a typological view of racial heredity, although he did not understand the causes for the persistence of the type in Quetelet's terms. Galton was much concerned with the reversion to what he considered to be the "ancestral type," as shown by his statistical reasoning; there was little room in this model for sudden change of the type itself, and so Galton appealed to discontinuous variation through the sudden creation of sports to explain species change. Unlike Quetelet, however, who seemed to believe that an analysis of racial type could be made by a comparison of means alone (*Letters on Probability,* p. 96), Galton was clear that all the characteristics of the distribution, including its variation, had to be considered.

43. Galton, *Memories,* p. 302. Another element in Galton's final discovery of the correlation coefficient was his interest in the 1880s in anthropometric statistics as utilized by Bertillion for criminal identification. This is discussed in Pearson, *Francis Galton* 2:383–4.

44. Galton, *Memories,* p. 305.

9 Alfred Marshall and the Development of Economics as a Science

H. SCOTT GORDON
Indiana University and Queen's University (Canada)

INTRODUCTION

In the evolution of the structure of modern economic theory, the last quarter of the nineteenth century stands out as a remarkable period—a transformation era which separates modern "Economics" from the "Political Economy" of earlier times. Some historians of economics have emphasized this difference even to the point of declaring that economics as a science only began in the 1870s, while others have insisted on the continuity of its development and regard the new as a natural outgrowth from the old. Alfred Marshall was himself of the latter view, not only because he saw his own work in economics as a continuation of that of David Ricardo and John Stuart Mill, but because he believed in continuity in itself, as constituting virtually a metaphysical principle of nature. Marshall's *Principles of Economics* (1890), which became the main supporting beam of the edifice of modern economic theory, carried on its title page the motto "Natura non facit saltum" (nature does not make jumps), which had been a favorite mental touchstone for Charles Darwin. Leibniz had apparently been fond of it too, and there is at least some curiosity in the fact that Marshallian (and modern) economics exhibits certain affinities to Leibniz's conception of reality as a "monadological" network of inter-reflections.

Nonetheless, despite Marshall's own predilection for ancestor worship, and despite the fact that many connections between the old "classical" political economy and the new economics of Marshall and his contemporaries may be traced, I lean personally to a discontinuity view—that something quite new and different began to emerge in economic theory in the 1870s. Marshall was the most important figure in the construction of what were essentially new foundations for the science. A modern economist who reads Adam Smith, Ricardo, the Mills, or Marx, for example, cannot avoid the feeling that he is visiting with the ancients, whereas Marshall, Wicksteed, Fisher, Pareto, Wicksell and the other great economists of the late nineteenth century, though outmoded in many of the specifics of their thought, still seem to belong to our own age in their general conceptions of economic theory and its relation to real economic processes.

Now, this appears to be leading up to an apology (or justification) for confining the scope of my paper to the last quarter of the nineteenth century in delineating Marshall's contribution to the main body of modern economic theory. However, only the first section of this paper will focus on issues which lie within the main corpus of modern economic theory. The other sections will take up some problems which were clearly in the forefront of Marshall's study of economic processes but which neither he nor his successors succeeded in penetrating effectively by means of the methodology of modern economics. In effect, I hope to throw some light on the methodology of late and early nineteenth-century economics by considering some of its weaknesses as well as its strengths.

I. MARSHALL AS SMITHIAN: "ECONOMIC MAN" AND THE OPERATION OF AN "INVISIBLE HAND"

A central methodological pillar of modern economic theory, which Marshall and his contemporaries adopted from their classical predecessors, is the concept of "economic man." The idea is usually identified with Adam Smith, though in fact it is present, implicitly or explicitly, in most of the pre-Smithian economic writings, in which an effort is made to be analytical about economic processes, rather than merely hortatory about personal economic behavior. The concept of "economic man" was largely responsible for the hatred of the early nineteenth-century economists by the philosophical humanists and romantics, who not only abhorred the changes then taking place in the real economy but poured scorn upon the pretensions of "Political Economy" to the stature of science, viewing it instead as a mere rationalization for the rapacious greed of industry and commerce executed by a mode of reasoning which was a blasphemy upon the noblest creation of the divine spirit. That view of economics, in strong or mild terms, has had a continuous history from the days of Dickens and Carlyle down to the present. It is, however, based upon some serious errors concerning the methodology of economic theory and misapprehensions about the content and implications of the concept of "economic man." To make headway in understanding the position of economics in the development of scientific methodology in the nineteenth century, it is necessary to elucidate the conception of "economic man" in a fashion that is free from the misconceptions that have been attached to it in widespread common opinion from the early nineteenth century to the present.

The first point I should like to make has to do with the relation between the concept of economic man and the use, in economic analysis, of collective entities such as classes, nations, races, etc. Any reader of the classical economists will appreciate the extent to which they took it for granted that nations are distinct entities and that, within nations, the population was essentially heterogeneous in socio-economic terms, being composed of distinct and definite classes, or "orders of men" as Adam Smith would say. The early political economists had no doubt that classes were "real" features of society—they did not regard them as methodological fictions invented for analytical purposes. The important issue is whether the classes are to be regarded as more real (so far as economic phenomena are concerned) than the individuals of which they are composed. Are classes simply generic classifications, the bases of which can be elucidated in straightforward empirical terms (such as, for example, a statistical demonstration that the variance, in some quality, within the membership of the classes is significantly less than that between the classes); or are classes to be regarded as super-organisms or some other type of wholes, the members of which are only parts? Is the economic behavior of a given class of men relatively homogeneous because the individuals are operating under the same economic and social conditions and constraints and therefore will, on account of their rationality, make similar decisions; or is it homogeneous because it is the class, *eo ipso,* which is the acting entity, individual behavior being essentially "epiphenomenal"?

The significance of the Smithian concept of "economic man" is that it represents a definite point of view on this crucial philosophical issue. By means of it the early economists established the methodological principle that, despite the reality of socio-economic classes, the primary entities which are conceived to operate in economic processes are individuals, rather than social collectivities of any type. The early economists were not unconscious of the weight of custom and other forms of constraints upon individual behavior; those of the classical era were inclined to believe that freedom of individual action was decidedly larger for the middle classes than for either the upper or the lower orders, but this was merely a matter of degree; all men, of all orders, were construed, for methodological purposes, to act individually as economic men rather than as representatives of classes.

The concept of "economic man" therefore reflects an important methodological principle, without which it is not possible to conceive of modern economic theory in its present configuration—the principle of "methodological individualism." It asserts that an analysis of economic phenomena, to be satisfactory, must trace the phenomena back to the actions of individuals, whose decisions of action are presumed to be rational in the sense that they are aimed at specific and stable objectives, and involve an appreciation of the real connections that exist between means and ends. To be sure, there is today recognition of the existence of important institutional entities which act as decision-making centers—notably *firms* and *governments*—but the "economic point of view" concerning them has been to resist explanations of their behavior that run in sociological, social-psychological, or political-philosophical terms, in favor of explanations which view their collective activity as a proxy for, or a reflection of, the rational economic behavior of their component (groups of) individuals. This character of modern economics, which has, especially since the late nineteenth century, been subject to intense criticism by economists themselves, bears historical witness to the early nineteenth-century

British origins of the science, but it also bears methodological witness to the inability of economic theory (thus far at least) to make much headway in any other direction.

This basic feature of economic theory was badly masked by Malthus' *Essay on Population* (1798), which many nineteenth-century writers took as a prototype of economic thinking in general. Malthus opened his argument by asserting two "postulata," which he meant to be taken as fixed "laws of nature": that man must eat and equally necessarily, must cohabit sexually, and therefore he will procreate up to some external constraint. Malthus' man is not a creature of will and choice (at least in the first edition of the *Essay*), he is, in fact, merely a gastro-intestinal tract and a reproductive system, which operate according to natural necessities. Many nineteenth-century interpreters of economics viewed the Smithian conception of "economic man" as only a slightly enlarged version of Malthus' food and sex machine, but this was a fundamental misinterpretation.

The central issue is, of course, that of determinism. Without entering upon a discussion of that hoary philosophical issue, it is necessary to note that the mainstream of nineteenth-century economic theory was a rejection of determinism. This was the real meaning of the Smithian concept of "economic man," as I have tried to indicate, and it was also the implication of another conception of economic theory that was equally an object of romantic abhorrence: the conception of the economic world as inherently one of *scarcity*. This is not the same thing as saying that it is inherently one of poverty (which is what the strict form of the Malthusian argument amounted to). Its real import is that human existence is infused with the necessity of making *choices*. Scarcity and choice go inextricably together and choice necessarily involves three conditions: having ends; being able to relate means to them in a rational and empirical way; and being free to act on the connections perceived. Paradoxically, man is unfree at both extremes of the economic spectrum—when his life is so poor that all he can do is live on the edge of subsistence (i.e. a Malthusian world), and when there is such great plenty that no choices are necessary (i.e. in "heaven").[1]

It is important to note in this connection that "the economic point of view" says nothing about the nature of the preferences that are presumed to underlie the individual acts of choice. Economic theory does not require that an individual should prefer commodities to culture, or even that he put his own material welfare before that of others. People may prefer Beethoven to beer, or social welfare to personal wealth, or justice to power, without the basic model of economic theory being impaired. All that is required for the working of the "logic of rational choice," as economic theory has been called, is that the structure of people's preferences be such that a collection of diverse good things is regarded as being superior to an exclusive diet of only one good thing. If preferences are of such a structure, then advantageous trading becomes possible, that is, the giving up of a bit of one thing in order to obtain a bit of another, and this trading (including the act of producing, which is a form of trading or exchanging) is the essence of the economic system.[2] When we find Alfred Marshall saying that economics deals with "man's measurable motives" it is neither a vague nor a limited definition of the science. Marshall was trying to avoid the label of narrow utilitarianism which had been attached, with some justice, to the earlier economists. He freely recognized that man is a varied and complex being. But in pursuing these unsimple objectives man is a rational creature, and, insofar as his motives lead to action, his acts are a revelation of his preferences. When they fix themselves in something objectively observable, or "measurable," such as market prices or quantities of goods produced or consumed, it is then possible to analyze the preferences and the economic phenomena to which their exercise gives rise.

A late Victorian–Edwardian representative of this point of view, clearer than Marshall, was Phillip Henry Wicksteed, Unitarian cleric, medievalist, authority on Dante. Wicksteed taught himself economics and then taught himself the differential calculus when he perceived that many of the central problems of economic theory consisted of defining the conditions of maxima or minima. In Wicksteed's view,[3] economics was simply an application of a general organon of reason, applicable to all cases where choice is necessary, whether the alternatives being considered were utilitarian and material, political, philosophical or spiritual. But Wicksteed is only superor to Marshall on this ground because he is the more straightforward and less cautious (and perhaps less perceptive of difficulties) of the two. Basically, Marshall held the same view: the "economic man" of Adam Smith was a man of reason and freedom.

The most important technical issue that is raised by the principle of methodological individualism, so far as economic theory is concerned, is the relationship between these acts of individual choice and the organization of economic life as a whole. Adam Smith began his study of the *Nature and Causes of the Wealth of Nations* with an examination of the division of labor, recognizing that most of a man's wants and needs are met by the economic activity of others rather than by his own direct efforts. How is this individual activity knitted together? What makes it into an orderly web instead of a chaotic tangle? Here again poor Adam Smith has been maligned by history, credited with the belief that there was an "invisible hand" which formed a perfect structure if left to do its work without interference. Such an interpretation considers Adam Smith to be a philosophical or metaphysical "monist" holding the view that there is a natural harmony in human ends and that all good or virtuous things are linked together in a perfect seamless whole. That such a view is to be found in philosophical literature one cannot doubt; that it is to be found in Adam Smith can only be held if one does not read him at all, or only with impermeable preconceptions.[4] The term "invisible hand" is Smith's own and it invites misconceptions as a result of the verbal resonances of the adjectional term—"invisible" seems to denote a power belonging to the spiritual rather than the material world and from this it is a short step, or stumble, to "divine" and "perfect." But nothing of this sort is involved in Smith's view of the economic order. The organizing mechanism of the economy is not "invisible" in any occult or metaphysical sense but only in the sense that one cannot point to any specific agent, agency, or organization which functions to articulate the activities of butcher, baker, and candlemaker to one another by means of a deliberate set of acts and orders. What functions is only an analogue of such a concrete entity—the set of markets in which suppliers and demanders meet to engage in acts of exchange. Smith's perception is only suggestive. He was aware that economic organization takes place via markets and he knew that prices and profit opportunities are central features of it, but he did no more than sketch in the broad outlines. In a way of speaking, the task of economic theory ever since Adam Smith has been that of rendering the operations of the invisible hand intelligible (and by doing so, to discover how it needs to be strengthened or supplemented). It is especially in this respect that the third quarter of the nineteenth century stands out as a remarkable era in the history of economics. It was then that the basic anatomy and physiology of the invisible hand—the market mechanism—was satisfactorily delineated. In Alfred Marshall's *Principles of Economics* (1890) it is no longer merely suggestive and there are no longer the least grounds for interpreting (i.e. misinterpreting) it as a metaphysical monism or anything of the sort. The basic forces of the economy are the wants and desires of individuals, operating within the institutional and natural constraints of their social and physical environment. These desires come together in markets and the prices there determined react back upon the individual actors and modify their behavior into mutually compatible adjustment. J. M. Keynes, Marshall's most distinguished pupil, chose an apt simile in his description of Marshall's economics in his obituary notice: "The general idea, underlying the proposition that Value is determined at the equilibrium point of Demand and Supply, was extended so as to discover a whole Copernican system, by which all the elements of the economic universe are kept in their places by mutual counterpoise and interaction."[5]

In the history of economic theory this conception of the economy is most commonly associated with the name of Leon Walras, founder of the "School of Lausanne," whose *Eléments d'Economie Politique Pure* (1874) is usually taken as the first clear statement of the theory of "general equilibrium." The publication of Marshall's great treatise took place sixteen years later, but the same comprehensive conception had been independently attained by him and was a central methodological feature of the "Cambridge School" of economics which he founded. The approach of Marshallian economics may be viewed thus: initially one may ask the question how the actors in the economic process, i.e., producers (or "firms") and consumers (or "households"), accommodate themselves to the environmental constraints in which they find themselves, for example, how a firm "optimizes" between the markets where factors of production are available at prices and the markets where goods may be sold at prices. However, one cannot stop with this set of solutions, because many of the "conditions of the environment" by which producing firms and consuming households are constrained are themselves the effects of their own economic activity, in the aggregate if not individually. The theoretical consequence of this is that the economy must be conceived as a matrix of simultaneous equations, which is what Keynes meant in describing Marshall's economics as "a whole Copernician system." The main theoretical focus of the Lausanne and Cambridge schools of economics was the delineation of this system: to discover whether it is determinate, whether it has unique or multiple solutions, whether the dynamic processes which drive it are compatible with the conception of equilibrium, whether the equilibrium has qualities of stability, and various other problems springing from the general conception and methodology which it represents. In addition, much attention was paid to defects of the invisible hand so depicted, and the study of such defects makes up the body of modern "welfare economics" which was established as a major branch of formal economic theory by A. C. Pigou, who was Marshall's successor at Cambridge, and by Vilfredo Pareto, who was Walras's successor at Lausanne.

II. MARSHALL AS EVOLUTIONIST: "THE MECCA OF THE ECONOMIST LIES IN ECONOMIC BIOLOGY"

The idea of general equilibrium, especially when illustrated by metaphors drawn from astronomy, seems to be a mechanistic conception of economic processes, leaving little or no room for the essentially "organic" elements of growth and evolutionary change. One would not expect social scientists to be fully satisfied by any purely mechanistic theory, and especially so those who lived in the latter half of the nineteenth century, in an intellectual and scientific climate that was so much affected as it was by Darwin. Almost all of the leading economists of the time expressed their sense of inadequacy with the mechanistic formulations of the developing body of economic theory: some of them even rejected it outright on this account and turned to descriptive historical or institutional research; others simply noted the defect and went on with the intellectually exciting work of amplifying the equilibrium model; only a few devoted much time and energy to organic and evolutionary problems while continuing to work along the (soon to be orthodox) lines of the equilibrium model. Alfred Marshall was one of the latter. In the Preface to the first edition of the *Principles* he claimed that it was the awareness of the principle of *continuity* that gave the book any special character it might possess and noted that the principle applied to historical processes as well as to other aspects of human life and thought. The influence of biology in this respect, and of the philosophy of history, Marshall asserted (referring to Herbert Spencer in regard to the former and Hegel in regard to the latter) to have "affected, more than any other, the substance of the views expressed in the present book." In an article in the *Economic Journal* of March 1898, Marshall issued the methodological declaration that "the Mecca of the economist lies in economic biology" and incorporated the epigram in the prefaces to all further editions of the *Principles*. He seemed in effect to be anticipating, and recommending, a "scientific revolution" in the meaning of that phrase as used recently in T. S. Kuhn's much-discussed book, *The Structure of Scientific Revolutions*. He appeared to be advocating a change in the basic "paradigm" of economic theory—from mechanism to organism. Marshall was not a methodological revolutionary, but became one of the modern pillars of the established orthodoxy in economics. He was clearly very much attracted to the prospects of an organic methodology, but in his own work he found it impossible to make a leap of this order. Moreover, he eschewed leaps, in intellectual method as well as in real life. He wished to graft biological conceptions onto the main stem of economic thought, hoping that in time the whole branch would become organic, but if the grafts would not take, which they did not, he had no intention of abandoning the parent root of the science.

The organismic theory of society has a long history in political theory going back at least to Plato, and may be found expressed by some writers as virtually a literal, rather than a merely metaphorical, interpretation of society and the state (e.g., by Saint Paul, John of Salisbury, Marsiglio of Padua, Nicholas of Cusa; and in the nineteenth century, e.g., by such writers as E. Forset in *A Comparative*

Discourse of Bodies Natural and Political and A. Schaffle in *Bau und Leben des sozialen Korpers*), but very little use of the conception is made by the great economists. Marshall's ideas along these lines are not coherently developed, and reflect more an awareness of the weaknesses and lacunae of the mechanistic paradigm in the methodology of economics than a positive vision of the nature of "Mecca." An examination of this aspect of Marshall's thought therefore reveals some important aspects of the established methodology of economics by delineating its limitations rather than its powers.

Although the paradigm of *organism* can be employed as the basic conception upon which a coherent epistemology may be constructed, it is doubtful that such an integrated complex of ideas was grasped by Marshall or by others who employed similar phraseology. The use of the term often represented little more than the fact that during the later nineteenth century, the science of biology seemed to challenge the mechanical conceptions of physics and astronomy and open up new vistas of epistemological and philosophical development. To many a late-century scholar, the term "biological science" signified as much by what it was not supposed to be as by what it was, since a large part of the methodological interest resided in the contrast which numerous writers drew between the mechanistic view or the natural world and the organic view. The former, represented most clearly by physics and astronomy, pictured the universe as a group of bodies whose relations to one another were determined by a limited number of general laws. Any particular system of these bodies, whether on a microscopic or a macroscopic level, was in the nature of an equilibrium system. External influences could cause movement and change within the system until a new equilibrium, representing the compatibility of the fundamental laws, was again established. Such a methodology in effect draws a sharp distinction between the *forces* of change and the *rules* of change. The former are external to the particular system under examination while the latter are internal to it. The rules of change are implicit in those laws which constitute the fundamental form of the system. They establish a set of relationships which, by defining the conditions of equilibrium or rest, also indicate what will be the result of changes in the magnitudes of any of the parameters of the system. But such changes are of an "external" nature, i.e., external to the system sketched by this particular set of equations, and the process of change is simply one of adjustment on the part of the system to exogenous forces. A model of this sort can be termed "mechanical" because it is illustrated by simple cases of physical mechanics, such as the motion of a pendulum, the action of pulleys and levers, etc. By contrast, the organic view claims that a fundamental law of the universe is the persistence of alteration and change, and that it is methodologically invalid to separate the forces of change and the rules of change. Accordingly, an analysis of a system in terms of the equilibrium relations of its parts is, at best, only preliminary. The significant feature of the organic conception is that the system (or organism) is conceived to possess, *within itself,* power to change and vary. Change is, consequently, not merely the adjustment of the system to a new force or an altered parameter, it is endogenous to the organism itself.

Marshall did not eschew mechanistic conceptions and illustrations. Many of his most significant contributions to economic analysis were mechanical in form, and indeed, the mechanical aspects of Marshall's analysis were so well machined, and so carefully articulated, that they constituted, by themselves, a fully integrated view of the economic world—again the "Copernican system" to which Keynes referred. In Marshall's own estimation, however, the construction of such a system was only the beginning of theoretical knowledge about the economy. Mecca was to be reached via this station but it was located far beyond it. Marshall's frequently expressed desire to get at the "causes of causes" reflected a feeling, which possessed some justification, that a mechanistic methodology tended to rest content with the discovery of proximate causes, or, at least, that it did not seek to penetrate beyond the realm of clearly seen and provable causal connections. His aim was to trace economic events back to such basic factors as the effects of geography on human character, the different temperaments and personalities of different races, and so forth. Many of the views which Marshall advanced in this connection strike one today as entirely naive, but an important point would be lost if one dismissed those passages in his writings as the curious opinions of a Victorian Englishman who allowed himself to be too much influenced by German romantic philosophy. Marshall's specifics may well have been wide of the mark, but they have a broader historical interest, representing as they do his attempt to come to grips with the "causes of causes" in economic evolution.

When he spoke of "economic biology" Marshall did not mean merely to distinguish dynamics from statics, for dynamics was also a mechanistic conception, and when he treated the problem of dynamics in the *Principles* it was in the form of an application of the dimension of time to the theory of mechanical statics. The study of both statics and dynamics was, in his words, but "a necessary introduction to a more philosophic treatment of society as an organism." An organic system was not merely one that moved but one that had, in itself, power of movement. On the other hand, it would be a mistake to infer that Marshall viewed society as "organic" only to the extent that it changed and progressed. Even a static society was conceived to be an essentially organic one, the true nature of equilibrium being not a state of rest, but "a balancing of the forces of growth and decay." The analysis of the economy as a mechanical balance, whether static or dynamic, was simply an analytical device, useful in approaching the more difficult problem of organic balance,[6] which was, in turn, only a step on the way to an understanding of the nature of organic evolution.

As long as one discusses these questions in such a general way as this no substantive issues of great importance seem to be involved. One economist might simply prefer to think of the economy as an astronomer might, and another might prefer the mind-set of a dynamic ecologist, but the basic theory and its applications would be essentially the same. There is, however, a vital point at issue. Economic theory, built largely on Marshall's own contributions, investigates the analytics of the problem of allocating the scarce means which man has at his disposal so as best to serve his given ends or objectives. Economic theory need say nothing about the normative or ethical quality of these ends and it is not even formally necessary to specify whose ends they are, but it *is* necessary to assume that they are "given" in the sense of being independent of the economic process itself. If this is not so, then the ends, and the means taken to satisfy them, are not independent of one another, and it is impossible to discover a determinate solution which one could describe as an "optimal" use of the scarce resources. This is the central issue of Marshall's organic view. It is only a very primitive society, he felt, to which an economic theory which is based on the exogenous nature of ends is applicable. Only a society that is very poor can be regarded as having simple or "animal" needs or ends; all other societies have ends which are sophisticated, and this feature of human wants becomes more important the more advanced the society. This alone would create no insoluble problems, but in Marshall's view, the wants of civilized man are not only sophisticated, but they are in significant part generated by economic processes themselves and are therefore endogenous to the system that one is theorizing about. The consequence is that man changes in his character and personality as he engages in economic activities undertaken to satisfy his wants. Consequently, there is no determinate equilibrium. The goal changes as man moves towards it and the tracks that human societies have made on the sands of history are not describable as efforts to reach a definite point, for the point changes with every movement, and on account of that movement. This comes close to laying an axe to the hope of a "scientific" methodology in the study of social phenomena, since, to a considerable degree, it denies that social phenomena come under the postulate of the uniformity of nature which is a necessary constituent of any science.

Marshall speaks often in his various writings of the great significance of the fact that man is the creation of social circumstances, not a constant to whose ends social institutions are adapted as utilitarian instrumentalities. He seems sometimes to echo J. S. Mill's view in *The System of Logic* that a general science of society can be founded on knowledge of "ethology"—the study of the formation of human character. But "ethology" in this sense was not to be, and Marshall was not one to lose firm hold of the solid methodological achievements of rationalism to follow the siren song of philosophical romanticism. His line of thought in "biological economics" had led him to a dilemma, to all solutions of which he was temperamentally and philosophically averse.

One can appreciate the strong appeal of solutions via the postulation of a teleology, or of some conception of "laws of history" which regards the process of evolution as being directed upon some metaphysical plane, but Marshall had no taste at all for speculations of that sort even though he admired some of those who were daring enough to make them. He devoted many years to historical research, and became an accomplished economic historian, but he was unwilling to follow Marx or Hegel into the postulation of historical "laws." There are some suggestions in his writings of the sort of ideas which led to the metaphysical models of such writers as Bergson, Alexander, and Smuts, but he was also quite unwilling to go very far in the same speculative directions himself. Any of these philosophical routes would have required him to abandon the methodological corpus of economic theory, which he would not do.

The net result is that Marshall only succeeded in raising the problem. Successor Marshallians, notably Frank Knight, have tackled it since, but no one has succeeded in advancing economic theory towards the Mecca that Marshall postulated. The mechanistic methodolgy of nineteenth-century economics proved to be a powerful "organon" even though it was inherently limited in what it could do. Economists who were more impressed by the limitations of the mechanistic mode than by its powers, in almost all cases turned to sociology or history, or to descriptive studies of economic institutions, and abandoned economic theory rather than revising it.

III. ECONOMIC THEORY AND ECONOMIC HISTORY:
J. A. SCHUMPETER AS MARSHALLIAN EVOLUTIONIST

The failure of Marshall, and that of his successors who took up the problem, to provide a theory of society in which human wants or ends are endogenous, made it impossible for economics to develop in a rigorous fashion any theory of economic history in the large. The methodology of equilibrium mechanics can be applied to particular industries or to particular events and has been made to yield a great deal of light thereby, but the grand sweep of economic development—large phenomena like the industrial revolution, the rise and evolution of the market economy and capitalism—eludes the net that the methodology has been able to cast. Economic history, in this grand scope at least, lies outside the economic model. A substantial literature of the "theory of economic growth" exists, which has been developed from the main stem of classical and neo-classical economics, but this contains purely formal scenarios of logically possible growth processes rather than theories of economic development as cognitive models by which the course of actual development may be delineated and analyzed. However, one of the earliest efforts to construct an analysis of the process of economic development based on Marshallian economics—that of Joseph A. Schumpeter—was more truly an effort to produce a theory of economic history in the general sense. Evolutionist conceptions are evident in Schumpeter's thought, but, like Marshall, he preferred the solid methodological power of the mechanistic model to the uncertain "insights" of organicist metaphysics, and he did not succeed in breathing evolutionary life into the economic approach to history. In this section I will discuss Schumpeter's theory, which was first advanced in his *Theory of Economic Development* (1911). The structure of this theory and its limitations assist one in perceiving the methodological character of economics as it was established by the great nineteenth-century writers.

Schumpeter's method in advancing his theory of economic development is to describe first the "circular flow of economic life" in a state of equilibrium. This economy is one that is experiencing no development. It repeats the same processes over and over, undergoes neither expansion nor contraction in aggregate output, and the incomes of the two primary productive factors—"land" and "labor"—exhaust this output. This description is "unrealistic" in that Schumpeter does not believe an economy ever to be in such a state of actuality; the economy may approach such an equilibrium during the depression phase of the business cycle, and, to some extent, the circular flow description is "real," but its primary function is to act as an analytical fiction that serves to isolate the phenomenon of development.[7]

In the *Theory of Economic Development,* Schumpeter repeatedly advances the view that it is his objective not to explain the ultimate causes of development, but to describe the mechanism by means of which the developmental process takes place. He uses the term "mechanism" often in this sense in the book. In the opening pages he argues that the task of economic theory is finished when causal analysis has been pushed to factors that are outside economics, that is, when "we ground upon a non-economic bottom" as he expresses it. This seems to suggest that the basic forces of economic change are conceived to lie outside the model. On the other hand, Schumpeter explicitly regarded his theory of economic development in a contrary way: "By development," he says, "we shall understand only such changes in economic life as are not forced upon it from without but arise by its own initiative from within."[8] This is the issue I wish here to focus on. Is Schumpeter's theory an endogenous or exogenous theory of development? Does the mechanism work by itself or does it require the intrusion of external factors?

The resolution of the issue depends upon the nature of Schumpeter's concept of "entrepreneurship" and specifically on whether this factor, which in his model is the basic dynamic force of economic change, is regarded as lying inside or outside of the model itself. Schumpeter's theory places it within, but if we examine it closely, the feat is accomplished by means of a conceptualization of entrepreneurship which renders the theory as a whole sterile.

The fundamental characteristic of entrepreneurial activity, from which all development springs, is that it consists in dealing with *novelty*. Schumpeter is not alone in realizing that this requires special philosophical and scientific attention. We may go back to René Descartes, who laid the foundations of mechanistic philosophy in the seventeenth century, and his conception of a dualistic ontology, differentiating man from all other elements in the universe, the human category being unique because of two factors: man's possession of a rational language, and his ability to deal with novelty. The first of these is really a derivative of the second, and it is man's innovative ability that throws serious doubt on a monolithic and mechanistic ontology. Contemporaneously with Schumpeter, Leonard T. Hobhouse was making the innovative faculty the basis of his great trilogy on social evolution[9] and Henri Bergson was attempting to find in it a philosophical alternative to the post-Cartesian monolithic mechanism of the natural sciences.[10] In view of such a widespread belief that the ability to deal with novelty is the most distinctive quality of the thinking human being, one might expect that Schumpeter, by conceptualizing entrepreneurship as innovation, aimed to advance an organic, non-deterministic, theory of economic development.

The entrepreneur of Schumpeter's analysis, however, undergoes a profound metamorphosis in the course of exposition. He starts as a human being particularly endowed with innovative ability, or the ability to handle novelty. He ends, not as an organism at all, but as a *function,* a functional part of a machine which is the "mechanism of economic development." The "entrepreneurial function," not the entrepreneur, is what operates in Schumpeter's theory of economic development. Thus Schumpeter emerges, in the analytic development of his views, as a thoroughgoing mechanist; his theory envisages the entrepreneurial function coming into play, not as a result of the free exercise of a *res cogitans,* but solely on account of the forces of economic development acting mechanically on a mathematical distribution of "entrepreneurial abilities."

The entrepreneurial function as such is carefully separated from all else in Schumpeter's discussion of the distinction between entrepreneur and "capitalist"—the latter being the provider of financial capital for new enterprises. Schumpeter recognizes that entrepreneur and capitalist may be combined in one human being, but he regards this as unlikely in real life, and he does not feel that a great departure from realism is introduced by considering capitalist and entrepreneur as completely separate entities. This separation in itself creates a logico-empirical contradiction, for it requires us to believe that entrepreneurs succeed in borrowing large sums of money from capitalists for new and untried schemes without risking any of their own funds. Such capitalists would either be weak-headed or *they would be exercising entrepreneurial perception themselves.* However, the principal objection that must be made to the strict separation of the entrepreneurial function is a purely logical one. A "function" must necessarily be defined in terms of what it does, and this turns Schumpeter's theory, as an endogenous model, into an unenlightening truism. Entrepreneurship being purely and simply the *function* of making changes, the theory succeeds in telling us that the function of making changes is important in an economy undergoing change!

Schumpeter's first book, published three years before the *Theory of Economic Development,* was a methodological treatise.[11] One of the arguments it contained was the desire to replace the concept of "cause" in scientific economics by "the more perfect concept of function."[12] Schumpeter apparently had in mind the idea that a system of mutual determination, as embodied in the Walrasian general equilibrium equations, is not a causal analysis in any simple sense. Each element is a functional part of the whole system rather than an element in a chain of cause and effect. Fritz Machlup states that Schumpeter later came to the opinion that this was a mistaken view of the scientific problem;[13] however, it is clear that an idea of this nature ruled his thought while the *Theory of Economic Development* was being written and he never succeeded in breaking free of it.

Another element of Schumpeter's methodology stressed by both Machlup and Haberler[14] is the concept of "methodological individualism" which was discussed earlier in this paper. Schumpeter correctly meant to assert the necessity of studying the motives, actions, and conduct of the human actors in the economic process. The principle of methodological individualism should have led to a less mechanistic theory than that which is found in the *Theory of Economic Development;*[15] to a large degree the basic weakness of Schumpeter's theory is due to his failure to follow out that principle. In the opening pages of *Development* it is clear that if he ever held such a principle, it had very early been superseded by a much narrower definition of economics.

In one short passage at the end of Chapter II, Schumpeter does consider a flesh and blood entrepreneur and talks about the motives of dominance, conquest, and creation which rule him, but he quickly draws back again, realizing that he has gone beyond the boundaries of economics. He ultimately goes in another direction: if entrepreneurship is merely a function, divorceable from those who perform it and brought into play by mechanical forces, one need only describe the mathematical distribution of the innovative qualities necessary to the performance of the function and all else can be simply deduced. Schumpeter's solution here is para-Darwinian. He assumes that innovative ability is a variation like any biological variation that exists in the progeny of a single progenitor. In the population as a whole, the distribution of innovative qualities follows the law of probability, which, as in biology, creates order out of the chaos of randomness. Ranging in degree according to the probability law then, innovative ability of the requisite degree is available for the hardest task of path-breaking novelty, for the lesser task of parallel exploration, and for the lowliest entrepreneurial task of imitation. Since opportunities for innovation always exist, according to Schumpeter, and since the whole range of entrepreneurial qualities necessary to take full advantage of them also exists, economic development is inevitable in a private enterprise society. Schumpeter's exploration of the entrepreneurial function and its distribution adds little to our understanding, for we emerge with the statement that if we assume a constant supply of developmental opportunities and a constant supply of the abilities necessary to turn them to account, economic development will occur. No one will deny the truth of such a statement but it is hard to credit it with the status of an explanatory theory.

The full mechanistic flavor of Schumpeter's approach is found in his explanation of the cyclical process (or rather, the *cumulative* character of the phases of the cycle, for the turning points are not well provided in his theory). The cumulative character of the boom is solely due to the automatic action of development itself upon the probability distribution of entrepreneurial abilities. Once the supreme innovation is made (and why and how this comes about is never adequately explained), opportunity is provided for the entrepreneurs of lesser innovative ability, whose action in turn opens the way for lesser entrepreneurs, and so on. The developmental process is itself a valve control which releases entrepreneurial activity in greater or lesser quantity depending upon the speed and direction of the movement. Nowhere in this process do we encounter a thinking being. The most uniquely human quality, which has given pause even to the most mechanistic philosophers, the power of innovation, is converted into a species of physical energy, mathematically distributed and mechanically released. The *res cogitans* becomes merely another example of the probability law, and human life becomes a behavioristic element of a machine universe. It is easily understandable why Schumpeter's cycle theory can be adequately illustrated by a mechanical (pendulum) analogy as Frisch had done,[16] but this is precisely its weakness. If we believe that the entrepreneurial "function" must necessarily be carried out by a human entrepreneur, a theory of the "mechanism of economic development" may perform the important task of permitting one's knowledge of such factors to display its full significance. But such factors must, of course, remain exogenous to the model itself. Schumpeter's work was fruitful in that it pointed to the study of entrepreneurship for any understanding of the process of development; it stimulated sociologists and historians, and some economists, to cultivate this area; but it failed to provide a proper analytic vehicle for the interpretation of their findings, because, through aspiring to be a completed endogenous theory in itself, it left no room for the incorporation of any empirical knowledge about human entrepreneurs and human entrepreneurship. In the final analysis, all that Schumpeter's theory can say about Andrew Carnegie is that he was a kind of extreme biological variation that proved to be highly successful, but how and why that was the case remains as mysterious as ever.

CONCLUDING COMMENTS

Properly conceived, the prime task of the student of the methodology of science is to describe, explain and analyze the successful methodologies in our scientific experience, rather than to prescribe for scientists the manner in which they ought to proceed. It is only against the reference of actual success that one can be critical of weak and deficient methodologies. Anyone who issues a general disparagement of a methodology that has found wide favor with the working men of science over a long period of time exposes himself to the riposte that he do it better himself—a challenge that rarely evokes a response. Brendan Behan's remark that the literary critic is like a eunuch in a harem—knowing exceptionally well how the trick is done, and its many complexities and subtleties, but incapable of performing it himself—applies, *mutatis mutandis,* to the historian and philosopher of science equally well. (I have left unstated here what one means by "success" in science, an issue that is more complex and less certain for the social than for the physical and natural sciences, only because it cannot be discussed briefly.)

Within the boundaries of such rules, however, it is valid and proper to recognize that an accepted methodology may be limited in scope, there being problems of interest and importance that it has not been able to attack, and it may be legitimate to suggest that its limitations are "inherent" in its methodology.

Many of the constituents of modern economics were suggested and developed by the classical "political economists," from Adam Smith to Karl Marx, but fundamentally they marched to a different rhythm from that of the late nineteenth- and twentieth-century economists. The classical writers hoped to unravel the skein of history, to find the great central forces which move (determine?) the course of nations and empires. As one modern author has put it, they were engaged in the delineation of the "magnificent dynamics." By contrast, the task set by the successors of Walras and Marshall seems mean and petty, but it was their efforts to analyze the mechanics of markets that produced the problem-solving economics we now possess. Modern economics can say very little about the laws of historical evolution—Marshall's "Mecca" was more of a mirage than an insight—but it can say a great deal that is relevant to immediate issues that are focal points of social concern and public policy. In the pragmatic test, it is the latter that deserves man's gratitude. Clio seems to be infinitely imaginative and refuses to allow us to bind her to law. As private and social beings we can do very little about "history," but we can do a great deal about the problems of this year and even the next, and that is well worth doing. Perhaps, indeed, that is all that "history" is.

NOTES

1. Actually, it is rather difficult to conceive of such a state even without any "economic" constraints. For there to be no necessity of choice, time as well as material goods would have to be unlimited, and this would require that each individual's command of time was not only infinite in extent, but also in some sense, "in depth," so that if one wished, more than one enjoyable thing could be done simultaneously.

2. W. D. Lamont, noting that the objective of an individual's economic activity need not be egoistic, suggests the use of the concept "the economic relation" rather than "the economic man," which he finds misleading on this score. See "The Concept of Welfare in Economics," The

Aristotelian Society, *Berkeley and Modern Problems,* Supplementary Volume 27, 1953.

3. See especially Wicksteed's *The Common Sense of Political Economy* (London: Macmillan, 1910).

4. In economic thought the issue here is that of "laissez-faire." Some doctrinaire advocates of laissez-faire existed in the era of classical economics but they were few, and without exception did not occupy leading positions in economic theory as such. The belief in a harmony of the natural order has had an interesting revival recently among certain biologists, professional and lay, and has come to focus on the science of ecology. Laissez-faire ecologists are however not full monists—the activity of man is regarded as an exception (the only exception?) to the natural harmony. The intellectual derivation of this branch of thought may be from Leibniz, but it is via Rousseau rather than Adam Smith.

5. J. M. Keynes "Alfred Marshall, 1842–1924" in A. C. Pigou, ed., *Memorials of Alfred Marshall* (London: Macmillan, 1924), p. 42.

6. Alfred Marshall, *Principles of Economics* (London: Macmillan, 1890). The opening passage of Book V of the *Principles,* which deals with the "General Relations of Demand, Supply and Value" reads as follows:

A business firm grows and attains great strength, and afterwards perhaps stagnates and decays; and at the turning point there is a balancing or equilibrium of the forces of life and decay: the latter part of Book IV has been chiefly occupied with such balancing of forces in the life and decay of a people, or of a method of industry or trading. And as we reach to the higher stages of our work, we shall need even more and more to think of economic forces as resembling those which make a young man grow in strength, till he reaches his prime; after which he gradually becomes stiff and inactive, till at last he sinks to make room for other and more vigorous life. But to prepare the way for this advanced study we want first to look at a simpler balancing of forces which corresponds rather to the mechanical equilibrium of a stone hanging by an elastic string, or of a number of balls resting against one another in a basin.

The contrast was drawn even more explicitly in the following passage:

Consider, for instance, the balancing of demand and supply. The words "balance" and "equilibrium" belong originally to the older science, physics; whence they have been taken over by biology. In the earlier stages of economics, we think of demand and supply as crude forces pressing against one another, and tending towards a mechanical equilibrium, but in the later stages, the balance or equilibrium is conceived not as between crude mechanical forces, but as between the organic forces of life and decay. ("Mechanical and Biological Analogies in Economics," *Memorials of Alfred Marshall,* p. 318.)

7. In the *Theory of Economic Development,* (trans. Redvers Opie [Cambridge: Harvard University Press, 1934]) Schumpeter does not directly discuss the methodological issue raised here. One receives the

impression that he is perhaps not quite prepared to accept this interpretation of the circular flow (e.g., see p. 245). In at least one other place, the early pages of his essay on "Social Classes" (*Imperialism and Social Classes,* ed. P. M. Sweezy [Oxford, 1951], pp. 137–221), Schumpeter supports this view unequivocally, at least so far as classificatory concepts are concerned. However, Schumpeter's methodological tolerance and eclecticism have been emphasized by a surprising number of the contributors to the memorial volume compiled on the occasion of his death (S. E. Harris, ed., *Schumpeter, Social Scientist* [Cambridge: Harvard University Press, 1951]). Of these, A. W. Marget is the most definite in interpreting Schumpeter's view of theory as a tool of cognition. It is significant however that Marget's references are to Schumpeter's comments on the work of Wesley Clair Mitchell (S. E. Harris, ed., *Schumpeter,* p. 67n) and it is perhaps not surprising that the aridness of Mitchell's strict empiricism should motivate Schumpeter to express himself more unequivocally than if he were making a positive statement on his own behalf. The monograph by R. V. Clemence and F. S. Doody on *The Schumpeterian System* (Cambridge: Addison-Wesley, 1950) discusses this question but is rather ambiguous on the point at issue (see, e.g. ch. V.).

8. *Theory of Economic Development,* p. 63. In a similar vein, Schumpeter rejects any evolutionist theory of continuous change through the accretion of small increments. Endogenous and discontinuous are the basic qualities his theory must possess: what we are about to consider is the kind of change arising from within the system *which so displaces its equilibrium point that the new one cannot be reached from the old by infinitesimal steps"* (Schumpeter's italics). (*Theory of Economic Development,* p. 67n.)

9. Leonard T. Hobhouse, *Mind in Evolution* (London, 1901); *Morals in Evolution* (London, 1906); and *Development and Purpose* (London, 1913).

10. Henri Bergson, *Creative Evolution,* trans. Arthur Mitchell (London, 1911).

11. Schumpeter, *Das Wesen und Hauptinhalt der theoretischen Nationalekonomie* (Leipzig, 1908).

12. See F. Machlup, "Schumpeter's Economic Methodology," S. E. Harris, ed., *Schumpeter,* p. 42.

13. Ibid., p. 97n.

14. "Joseph Alois Schumpeter, 1883–1950," ibid., p. 42.

15. Clemence and Doody's answer to the charge that Schumpeter's functional definition of innovation is tautological, is to interpret the concept along such humanistic lines. See, *The Schumpeterian System,* pp. 36–41. However the authors themselves recognize that Schumpeter gives little support for this defense in his own writings.

Schumpeter's "Social Classes" which was first published in German in 1927, contains a passage (Sweezy ed., *Imperialism*, pp. 154f) dealing with the Marxian theory of surplus value and investment which suggests that Schumpeter's investor is an effective individualistic actor, yet one who lives in a social environment that exerts its influence over him. That is to say, both individualist and behaviorist elements are at work. The suggestion however is not developed to any extent and one is moved to the conclusion that Schumpeter's criticism of Marx is made to rest on a foundation that his own *Theory of Economic Development* had already removed from consideration. (See also his critique of Marx in his *Capitalism, Socialism and Democracy* [London, 1943].)

16. Ragnar Frisch, "Propagation Problems and Impulse Problems in Dynamic Economics," *Economic Essays in Honour of Gustav Cassel* (London, 1933). The weakness of the turning points explanation is clearly shown by Frisch's pendulum analogy for no economic counterpart is given for the gravitational force which eventually overcomes the movement of the pendulum in any one direction. Haberler asserts that Schumpeter was never quite happy about Frisch's model ("Schumpeter's Theory of Interest," S. E. Harris, ed., *Schumpeter*), but remarks only that this was due to Schumpeter's general distrust of abstract mathematical formulations.

10 Grove Karl Gilbert and the Concept of "Hypothesis" in Late Nineteenth-Century Geology

DAVID B. KITTS
University of Oklahoma

I

In the preface of their widely used textbook of elementary geology, Leet and Judson say, "Originally, geology was essentially descriptive, a branch of natural history. But by the middle of the twentieth century it had developed into a full-fledged physical science making liberal use of chemistry, physics, and mathematics, and in turn contributing to their growth."[1] This kind of statement is not uncommon in the twentieth century, nor was it uncommon in the nineteenth. Geologists have been remarkably apologetic about their discipline although they have often expressed a hope, and sometimes even a promise, that things would get better. Many outside geology have expressed a similar view. The historians Basalla, Coleman and Kargon entitle the section on geology in their *Victorian Science,* "Geology Becomes a Science," clearly implying that before the late nineteenth century it had been something else.[2] It is far from clear what is being got at in these characterizations of geology, but a key to it all seems to lie in the term "descriptive." To describe something, according to my dictionary, is to represent it in words. There is apparently no restriction as to the subject of the representation. But usually, in science at least, we take the subject of a description to be a particular object or event rather than a universal condition. Thus it is quite natural to say that we describe a sunset and at least a little strange to say that we describe the laws of physics. When someone says that geology is descriptive, he usually means to convey that the body of geologic knowledge consists largely or wholly of statements about particular objects and events. An examination of the geologic literature of the nineteenth and twentieth centuries will reveal that geology is indeed overwhelmingly descriptive in this sense. I do not know why anyone should deny it or, having recognized it, should apologize for it. Geologists have always made a great deal of the fact that geology is historical. To say that geology is descriptive in the above sense is just another way of saying that it is historical. The historical concern of geologists is revealed in their preoccupation with spatio-temporal location or, to use a convenient term introduced by Simpson, with configurations.[3] Geologists, among all scientists, have elevated particulars to the status of significant items of knowledge, and in doing so have departed from an earlier tradition which held that the events which embody history are ephemeral, contingent and even unreal.

But there is an extreme version of the view that geology is descriptive. It is the view that geology is *merely* descriptive. It holds that geology consists largely, or wholly, of reports of direct observation. This is so patently absurd that one is bound to ask how anyone could believe it. I feel almost apologetic in noting at this point that accounts of the past and accounts of the present are formulated in the same descriptive terms. The descriptive vocabulary of geology does not, by itself, call attention to the inferential gap between assertions about the past and assertions about the present. This obvious feature of historical discourse has led geologists to an unjustified sense of familiarity with the past and has made it possible for some, both in and out of geology, to ignore the critical distinction between descriptions of the present and descriptions of the past. To focus upon this distinction is to focus upon the central methodological problem of geology. How do we get from the present to the past or, more generally, how do we derive and test singular descriptive statements?

There are different ways of getting at the past. We may, for example, simply make it up. Despite some claims to the contrary, accounts of the geologic past have never been so fanciful that it could be fairly said that they were made up. There have always been constraints placed upon what geologists are permitted to say about the past. They go from the present to the past by some rational process and it is for this reason that geology has been generally supposed to be a legitimate part of science.

II

The late nineteenth century was a period of inactivity in the philosophy and methodology of geology and so contrasted sharply with the first half of the century, which had seen an enormous amount of theological, philosophical and methodological discussion concerning geology. Perhaps philosophical interest in historical biology after 1860 had detracted from what by then might have seemed to be the less problematic issue of historical geology. Another factor, more important in North America than in Europe, was that geology had entered an intensely exploratory phase. As the exploration of the West proceeded, a vast area of well exposed rock became available for the careful examination of geologists. Geology became an activity in which running rivers and climbing mountains were at least as important as philosophical reflection. Coincident with this exploratory phase and, I think, related to it was a deep commitment to field work and meticulous observation as the only way to geological knowledge.

I offer no apology for choosing to discuss this period of poverty in geological philosophy and methodology. The early nineteenth century has received a great deal of attention from both historians and geologists, but the last half of the century has been little studied. My primary purpose in focusing upon this period is not simply to fill a gap in our knowledge of the history of geology, however. Rather I want to examine the establishment, during the last quarter of the nineteenth century, of a methodological view that persists to this day.

I have chosen to discuss Grove Karl Gilbert, who addressed himself to issues connected with the central methodological problem of geology. Gilbert was born in Rochester, New York in 1843 and graduated from the University of Rochester in 1862. He taught school for less than a year and then accepted a job as assistant in Ward's Cosmos Hall, which later became Ward's Natural Science Establishment. On the basis of the limited experience in geology acquired at Ward's, he was taken on as an assistant with J. S. Newberry's survey of Ohio in 1870. He spent 1871, 1872 and 1873 with the Wheeler Survey and in 1876 he joined the Powell Survey. He continued to work under Powell and King until 1879, when the United States Geological Survey was established. He joined the Survey and quickly rose to the rank of Chief Geologist, a position which he held until his death in 1918. If nothing else, this brief biographical sketch will serve to show that Gilbert was close to the heart of geological activity in the United States for nearly fifty years. He made significant contributions on a variety of subjects including geomorphology, structural geology, and economic geology.

Gilbert's principal methodological work, entitled "The inculcation of scientific method by example," was presented as the presidential address before the American Society of Naturalists in December 1885 and was published in the *American Journal of Science* the following year. Gilbert regarded it as an account of science as a whole, not of geology alone. His view of science is most explicitly stated in the following passage.

It is the province of research to discover the antecedents of phenomena. This is done with the aid of hypothesis. A phenomenon having been observed, or a group of phenomena having been established by empiric classification, the investigator invents an hypothesis in explanation. He then devises and applies a test of the validity of the hypothesis. If it does not stand the test he discards it and invents a new one. If it survives the test, he proceeds at once to devise a second test, and he thus continues until he finds an hypothesis that remains unscathed after all the tests his imagination can suggest.[4]

There is nothing particularly novel about Gilbert's view of science. At least it may not seem so at first glance. It differs, however, from the views held by physicists who were Gilbert's contemporaries. Despite substantial philosophical differences among late nineteenth-century physicists concerning the nature of science, most of them appear to have agreed that the central issues of methodology revolved about the concepts "law" and "theory." Gilbert does not mention these concepts in the quoted passage. Perhaps there is covert reference to them, in the terms "hypothesis" and "antecedent." Gilbert's use of these terms elsewhere, however, reveals that he uses them only to refer to particulars. "If the hypothetical antecedent is a familiar phenomenon," he says, "we compare its known or deduced consequences with A, and observe whether they agree or differ."[5] Gilbert's antecedents are not only logical antecedents; they are temporal antecedents. The term "hypothesis" also stands for initial conditions rather than for laws or principles. "Take first the hypothesis that the crust of the earth, floating on a molten nucleus, rose up in the region of the basin when its weight was locally diminished by the removal of the water of the lake."[6] So far as I have been able to determine, the use of the term "hypothesis" to stand for particular antecedent conditions is virtually universal among nineteenth- and twentieth-century geologists.

We must be careful not to make too much of this point. In a very important respect, geological science is a search for the temporal antecedents of phenomena. Antecedents are discovered with the aid of hypotheses, which consist, for Gilbert, of conjectures about specific initial conditions. Gilbert's account of science may simply reflect the geologist's preoccupation with the derivation and testing of singular descriptive statements. Most of us hold, of course, that the derivation and testing of singular descriptive statements requires the mediation of statements of general form, and we might suppose that Gilbert held this also. Let us examine his account more closely for evidence that beneath his primary concern for description there lies the recognition of a general or theoretical apparatus.

The term "law," which figures so importantly in discussions of the methodology of science of Gilbert's time, may suggest reference to general or theoretical notions.

It is known [Gilbert states] that the density of the earth's material increases downward, for the mean density of the earth, expressed in terms of the density of water, is about 5.5, while that of the upper portion of the crust is about 2.7. Nothing is known however of the law under which the density increases, and nothing is known as to the depth of the zone at which matter is sufficiently mobile to be moved beneath the Bonneville basin.[7]

Or take Gilbert's use of the term "postulate":

The verdict of the barometer was that the southerly shoreline was somewhat higher than the northerly, but the computations necessary to deduce it were not made until the mutual continuity of the two shore-lines had been ascertained by direct observation. The barometric mea-surement was therefore superseded as an answer to the original question, but it answered another which had not been asked, for it indicated that the ancient shore at one point had come to stand higher than that of an-other. The postulate of horizontality was thus overthrown.[8]

As I have already remarked, the term "hypothesis" and "ante-cedent" are sometimes used to designate specific initial conditions. The terms "law" and "postulate" almost never are. "Induction" may be used to designate an argument which leads to a general conclusion, or one which leads to a particular conclusion. In Gilbert's day, it was almost always used in connection with the "discovery" of laws and theories. In contrast, Gilbert used it in a wholly particular sense.

The condition of the interior of the earth is one of the great problems of our generation. Those who have approached it from the geologic side have based a broad induction on the structural phenomena of the visible portion of the earth's crust, and have reached the conclusion that the nucleus is mobile.[9]

This "broad induction" begins with particulars: "the structural phe-nomena of the visible portion of the earth's crust." But it does not lead to a general conclusion. It leads instead to a singular descriptive statement about the earth's interior. Consider another instance of Gilbert's use of "induction."

It is one of the great inductions of geology that as the ages roll by the surface of the earth rises and falls in a way that may be called undulatory. I do not refer to the anticlinal and synclinal flexures of strata, so conspicuous in some mountainous regions, but to broader and far greater flexures which are inconstant in position from period to pe-riod. By such undulations the Tertiary lake basins of the Far West were not only formed but were remodeled and rearranged many times. By such undulations the basin of Great Salt Lake was created.[10]

The conclusion that "the earth's crust rises and falls in a way that may be called undulatory" could be given a general interpretation. It does not figure as a general term in Gilbert's argument, however. Its most precise formulation would consist of a conjunction of singular de-scriptive statements, which is to say, a detailed historical account.[11]

However significant the above passages may be, in none of them does Gilbert directly address himself to the central methodological problem of geology. In one place he comes close to explicitly considering it.

> Given a phenomenon, A, whose antecedent we seek. First we ransack the memory for some different phenomenon, B, which has one or more features in common with A, and whose antecedent we know. Then we pass by analogy from the antecedent of B, to the hypothetical antecedent of A, solving the analogical proportion: as B is to A, so is the antecedent of B to the antecedent of A.[12]

I do not propose to discuss the logic of analogic arguments. I only wish to point out that Gilbert sees argument by analogy as a means of justifying a move from event to event by invoking a particular event rather than a principle or law.

III

Gilbert was not a philosopher, nor was he, like Hutton, a theoretical geologist of first rank. It is hardly surprising that he fails to give a complete account of geological inference. Theoretical statements do play a role in geology, and any satisfactory discussion of geological method must take them into account. Apparently Gilbert realizes this critical role although he nowhere states it. He uses the term "deduction," and he must understand that deductive arguments rest upon a premise of universal form. It is remarkable, however, that nowhere does Gilbert explicitly provide for a general foundation. Furthermore, by using terms for the singular and particular that stand for the general and theoretical in other contexts, he effectively cuts himself off from an analytical apparatus that could permit him to do so.

Gilbert's philosophical antecedents are obscure. Is he an inductivist or deductivist? It is almost absurd to ask. He seems to be operating outside any familiar philosophical tradition, and his frequent use of terms with a more or less established meaning within these familiar traditions does not conceal the fact. Nevertheless there are compelling reasons to take Gilbert seriously. Almost unanimously, contemporary geologists agree that he had something important to say, and many of them hold that he succeeded in getting at the critical problems of geological knowledge. "The Fabric of Geology," published in 1963 to commemorate the seventy-fifth anniversary of the Geological Society of America, is filled with allusions to Gilbert as the premier methodologist of modern geology. The least that we can hope to get from a study of Gilbert's notions about science is an insight into the methodological views of three generations of geologists. It is interesting to note that many of the geologists who extol the virtues of Gilbert profess to find nothing in contemporary philosophy of science which illuminates the problems of geological knowledge. How can we account for the strong appeal of Gilbert's methodology of geology where after all it counts the most, and that is among practicing geologists?

IV

Describing a prevailing attitude in the early eighteenth century, Lovejoy says, "The process of time brings no enrichment of the world's diversity; in a world which is the manifestation of eternal rationality, it could not conceivably do so. Yet it was in precisely the period when this implication of the old conceptions became most apparent that there began a reaction against it."[13] Geology was intimately involved in this reaction. Geology is the study of change and may, for some, become a study of the "enrichment of the world's diversity."

The reaction of which Lovejoy speaks raised grave philosophical questions which entailed serious methodological problems. If the process of time brings an enrichment of the world's diversity, then what limit is to be imposed upon that diversity? Some limit must be imposed. To impose no limit at all is to remove any constraint upon what may be supposed to have occurred in the past and, in effect, to remove the writing of geologic history from science, and indeed from rational discourse. For the geologist, this is no issue for idle philosophical speculation. He encounters it every time he performs an historical inference.

The classical debates of late eighteenth- and early nineteenth-century geology concern the character and magnitude of the restriction imposed by the immutable natural order upon the events of history. The solution presented by the two towering figures of this period, Hutton and Lyell, is a conservative one. To the question, "To what extent does the natural order permit an enrichment of the world's diversity in time?" their answer is, in effect, "To no great extent." In a recent article on Lyell's *Principles,* Rudwick states:

> Yet all these areas of emphasis are used in the strategy of the *Principles,* not as arguments of interest in themselves, but as tactical devices to be deployed in the service of a uniformitarian system of earth history.
> It is surely in Lyell's commitment to this system that we should look for a key to his reluctance to accept a progressionist view of the history of life or, ultimately, a transmutationist view of the mechanism of that progression. For it is ironic that the uniformity that later generations of geologists and biologists came to accept from Lyell was that of this actualistic methodology; they came to reject the uniformity of his steady-state system in favor of a developmental system much closer to that of his directionalist opponents.[14]

The directionalist view began to pervade geology in the late nineteenth century, and Gilbert's methodology should be viewed with this in mind. The major philosophical problem for the directionalist is the accommodation of real historical change, an increase in the world's diversity, with a real immutable order. Darwin provided a solution to the problem for the biological directionalists. The principles of Darwinian theory permit, indeed require, a highly directional account of history. Geologists had achieved no such dramatic solution to a difficult problem. It is not clear that they even saw it as a problem. Gilbert apparently did not. But there is a suggestion in the work of Gilbert, and in the work of other geologists in his time and later, that they felt that a rigid theoretical structure posed a threat to progressive history. They seem almost to have thought that an ultimate historical solution might be found in a method that permitted the jusification of inferences from event to event without the invocation of any general apparatus whatever. This judgment suffers at least two defects as an immediate explanation of Gilbert's particularization of method. First, there is nothing in Gilbert's work to indicate that he was consciously committed to directionalism or that he was aware of the increasing tendency of geologists to give directional accounts of history. More important is the fact that geologists do routinely invoke generalizations and theories despite the reluctance of geological methodologists to discuss the procedure, and geologists, including Gilbert, reveal that they are perfectly aware of it in their day to day work. To accept the view that Gilbert was attempting to enhance directionalist geology by finding a method that circumvented theory is to suppose that Gilbert simply overlooked the fact that he and his fellow geologists employed laws and theories as instruments of historical inference.

Geological observations and principles are formulated wholly within the context of a complex system of general preconceptions, so complex a system that one could not hope to identify all of its components with any reasonable effort. There is, however, a readily identifiable part of this system. It is what geologists consider to be the most fundamental and comprehensive principles contained in the physical theory of their day. These principles are regarded as wholly unproblematic for the purposes of geological inference.

It is neither logically nor empirically necessary that geologists operate within the conceptual framework of contemporary physical theory. The theory is not imposed upon geologists; geologists impose it upon themselves. Throughout the history of geology, physical theory has provided an immediately available inferential apparatus of great power and demonstrated utility. But this is not the primary reason for the geologists' decision to proceed under the umbrella of the fundamental theory of their time. Geologists do not regard these theories as optional formulations. The laws of chemistry and physics are more than "inference tickets." For most geologists they are "true" or "nearly true" statements about reality. In the mainstream of geology, the scientific character of the discipline is regarded as guaranteed by its demonstrable connection with physical theory. Physical theory is applied in geological inferences so directly and so obviously that it simply cannot be overlooked. Let us consider an example from Reade's discussion of overthrusts.

In attempts to unravel some of the weightier problems of geology it has lately been assumed that certain discordances of stratification are due to the thrusting of old rocks over those of a later geological age. Without in any way suggesting that the geology has in any particular instance been misread, I should like to point out the difficulties in accepting the explanation looked at from a dynamical point of view when applied on a scale that seems to ignore mechanical probabilities. Some of the enormous overthrusts postulated are estimated at figures approaching 100 miles. Have the authors considered that this means the movement of a solid block of rock or rocks of unknown length and thickness 100 miles over the underlying complex of newer rocks? If such a movement has ever taken place, would it not require an incalculable force to thrust the upper block over the lower, even with a clean fractured bed to move upon? Assuming that the block to be moved is the same length as the overthrust, the fracture-place would in area be $100 \times 100 = 10,000$ miles. I venture to think that no force applied in any of the mechanical ways known to us in Nature would move such a mass, be it ever so adjusted in thickness to the purpose, even if supplemented with a lubricant generously applied to the thrust-plane. These are thoughts that naturally occur to me, but as my mind is quite open to receive new ideas I shall be glad to know in what way the reasoning can be met by other thinkers.[15]

A geological inference leads to the conclusion that large masses of rocks have been laterally displaced over considerable distances and yet "no force applied in any of the mechanical ways known to us in Nature would move such a mass." "The mechanical ways known to us in Nature" are systematized in mechanics. A geological inference seems to require an event that the theory of mechanics forbids, and we have in the words of Hubbert and Rubey the "mechanical paradox of large overthrusts."[16] The paradox may be resolved by altering either the geological inference or the mechanical inference. In the mainstream of geological thought, no one hesitates over the choice. There is a remarkable consensus about what is more basic and fundamental and what is less so. "Basic" and "fundamental" here mean not only comprehensive but also inviolable. The geologist chooses to alter the geological inference in order to save mechanics. As soon as the decision to alter one inference to preserve the other is made, the means of achieving the alteration is suggested. In cases where the geological evidence is compelling, this resolution of the paradox will not consist of an out of hand rejection of the event, but rather of its alteration to bring it into accord with the fundamental theory. And in altering the event to conform to the theory, a geologist may discover something of great interest about the conditions of the past. Smoluchowski recognizes the paradox of overthrusts and adds a new dimension to it.

It is easy enough to calculate the force required to put a block of stone in sliding motion on a place bed, even if its length and breadth be 100 miles, and I do not think Mr. Mellard Reade meant to use the word "incalculable" in a literal sense. . . . Let us indicate the length, breadth, and height of the block by a, b, c, its weight per unit volume by w, the coefficient of sliding friction by e; then, according to well-known physical laws, a force $abcwe$ will be necessary to overcome the friction and to put the block into motion. Now, the pressure exerted by this force would be distributed over the cross-section ac; hence the pressure on the unit area will be equal to the weight of a column of height be. Putting $e = 0.15$ (friction of iron on iron), $b = 100$ miles, we get a height of 15 miles, while the breaking stress of granite corresponds to a height of only about 2 miles. Thus we may press the block with whatever force we like; we may eventually crush it, but we cannot succeed in moving it. The conclusion is quite striking, and so far we cannot but agree with Mr Reade's opinion.[17]

Smoluchowski goes on to suggest that the paradox may be resolved by hypothesizing either that the mass of rock moved along an inclined plane or that all or some part of the block was plastic rather than rigid. During the past four or five decades, geologists have invoked both gravitational sliding and reduced coefficients of friction for rocks involved to explain overthrusting. Others have felt that the independent evidence for slopes sufficient to result in major tectonic sliding of rocks with sufficiently low coefficients of friction is far from compelling. Hubbert and Rubey attempted another resolution of the paradox.

It therefore appears that, during periods of orogeny in the geologic past, which often have affected sedimentary sections many kilometers thick, the pressure in the water contained in large parts of these sediments must have been raised to, or approaching the limit of flotation of the overburden. This would greatly facilitate the deformation of the rocks involved, and the associated great overthrusts, whether motivated by a push from the rear or by a gravitational pull down an inclined surface, would no longer pose the enigma they have presented heretofore.[18]

Hubbert and Rubey believe they have an hypothesis that will resolve the paradox once and for all. They set out to support it by finding independent evidence for high fluid pressures in rocks involved in major overthrusts.[19]

The geologists who attempted to resolve the paradox of major overthrusts were directed at every juncture by their theoretical preconceptions. Their observations and descriptions and inferences were formulated in terms already imbued with theoretical significance. This is not to suggest that what they did was trivial or insignificant. They presented hypotheses and tested them, and in the process made significant contributions to geologic knowledge. But through it all there was one pervading hypothesis that was not subject to test at all. Mechanics was not tested against geological events. Geological events were tested against mechanics.

It is clear then that the theories of physics and chemistry are explicitly, obviously, and directly applied to problems in geology. How then can a geologist find Gilbert's account of geological method, which takes no account whatever of the role of theory, so appealing? The answer lies, I think, in the fact that geologists tend to see two distinct levels, or phases, of historical inference. Gilbert's method seems to account very well for the first phase, which consists of the initial step from the present to the past. If a geologist could be induced to admit that this step must be justified by laws and generalizations, he would be likely to claim that they are "self-evident" or "trivial" or even that they consist of "truisms." In the second phase of historical investigation, the events of the geologic past are "explained," or "interpreted," or "understood" in terms of physical theory.

Let us briefly reconsider the case of major alpine overthrusts. The inference which led to the conclusion that blocks of the earth's crust have been displaced laterally belongs to the first phase of historical investigation. A critical examination reveals that a number of principles must be adduced to justify this inference; to mention one, *the law of superposition.* This law asserts that in an undisturbed sequence of sedimentary rocks each bed is younger than the one below it and older than the one above it. For centuries geologists have been telling their students that the law of superposition is self-evident and have thereby done Steno, who formulated the law, and themselves, who use it everyday, a great injustice. It is self-evident, I suppose, that when objects are stacked up one after the other, the objects lower in the stack were put down earlier. It is not self-evident, however, that sedimentary rocks may be considered as members of the class of things that are stacked up one after the other. The justification for this assumption rests, not on its self-evident truth, but on an elaborate *theory* of sedimentary rocks which in turn rests upon physical and chemical theory. The attempt to explain the movement of blocks of the earth's crust in alpine thrusts belongs to the second phase of historical inference and invokes the explicit application of laws and theories which are, within the context of the inference, regarded as unproblematic.

The tendency to see two distinct levels of inference is quite characteristic of historical disciplines. It manifests itself in discussions of historical explanation which consider how historical events are explained and do not consider how historical events are obtained in the first place. It manifests itself also in discussions of the bearing of paleontological evidence upon the validity of evolutionary theories which do not take into account the extent to which the presupposition of those theories might condition that evidence.

I suggest that geologists find Gilbert's almost wholly particularized account of geological inference to be a satisfactory treatment of what might be called "primary historical inference." Geologists apparently do not regard this kind of inference as resting on theoretical, or even general principles. According to them, however, higher level historical inferences do.

In the mainstream of geology, the scientific character of the discipline has been assured by its demonstrable connections with physical theory. Geologists have not been able to explicate this connection in the concepts of philosophical analysis, but they have been able to secure and utilize the connection scientifically. The geology of the late nineteenth and twentieth centuries represents a triumph of the directional view of earth history. The highly directionalist concepts of continental drift and seafloor spreading have not been developed *in spite of* an immutable natural order, they have rather been developed in necessary connection with the most explicit expression of that order, contemporary physical theory. We might describe the geologists of the last one hundred years in Ryle's words: "They are like people who know their way about their own parish, but cannot construct or read a map of it, much less a map of the region or continent in which their parish lies."[20] The lack of such a map has not hampered geologists at all. Their account of a directional earth history within the context of a self-imposed rational order stands as one of the great intellectual achievements of our time.

NOTES

1. L. D. Leet and S. Judson, *Physical Geology* (New York, 1954), p. v.

2. G. Basalla, W. Coleman, and R. H. Kargon, eds., *Victorian Science* (New York, 1970).

3. G. G. Simpson, "Historical Science," in *The Fabric of Geology*, ed. C. C. Albritton (Reading, Mass., 1963), pp. 24–48.

4. G. K. Gilbert, "The Inculcation of Scientific Method by Example, with an Illustration Drawn from the Quarternary Geology of Utah," *American Journal of Science* 31, 3rd ser. (1886): 284–99.

5. Ibid., p. 287.

6. Ibid., p. 296. Gilbert's example for the "inculcation of the scientific method" is his ingenious attempt to explain the differences in elevation of the shoreline of the Pleistocene Lake Bonneville.

7. Ibid., p. 296.

8. Ibid., p. 291.

9. Ibid., p. 299.

10. Ibid., pp. 291–2.

11. "Induction" and "generalization" are frequently used by geologists, and sometimes by historians, to describe an argument that leads to conclusions which are "broad" rather than general, where "broad" stands for wide spatial and temporal extent. Thus Turner's "Frontier Hypothesis" is sometimes cited by historians as an historical generalization.

12. Gilbert, "Inculcation of Scientific Method," p. 287.

13. A. O. Lovejoy, *The Great Chain of Being* (New York, 1960), p. 284.

14. M. J. S. Rudwick, "The Strategy of Lyell's *Principles of Geology*," *Isis* 61 (1970): 32–3.

15. T. M. Reade, "The Mechanics of Overthrusts," *Geological Magazine,* new ser., dec. 5, 5 (1908): 518.

16. M. K. Hubbert and W. W. Rubey, "Role of Fluid Pressure in Mechanics of Overthrust Faulting," *Bulletin of the Geological Society of America* 70 (1959): 115–205.

17. M. S. Smoluchowski, "Some Remarks on the Mechanics of Overthrusts," *Geological Magazine,* new ser., dec. 5, 6 (1909) : 204–5.

18. Hubbert and Rubey, "Role of Fluid Pressure," p. 162.

19. Ibid., Pt. II.

20. G. Ryle, *The Concept of Mind* (New York, 1949), pp. 7–8.

11 Peirce and the Trivialization of the Self-Correcting Thesis*

LAURENS LAUDAN
University of Pittsburgh

If science lead us astray, more science will set us straight.[1]
—E. V. Davis (1914)

The aim of this paper is two-fold: first and primarily, to identify and to summarize the development of an important but hitherto unnoticed tradition in nineteenth-century methodological thought, and secondly, to suggest that certain aspects of the history of this tradition give us a new perspective from which to assess certain strains in contemporary philosophy of science. In part I below, I attempt to define this tradition, to document briefly its existence, and to note some features of its evolution. In part II, I sketchily attempt to indicate the manner in which this history may shed new light on some recent trends in inductive logic.

I

As the title of this essay suggests, the tradition that interests me is connected with the view of scientific inference as self-corrective, and the work of Charles Sanders Peirce looms large in the story.[2] It has been customary to see Peirce as the founder and first promulgator of the view that the methods of scientific inference are self-corrective.[3]

* I am very grateful to Adolf Grünbaum, Alex Michalos, John Nicholas, and Wesley Salmon for their helpful and generous comments on an earlier draft of this essay.

275

This historical claim is simply incorrect. The doctrine that scientific methods are self-corrective, that science in its development is inexorably moving closer to the truth by a process of successive approximation, has a pedigree extending back at least as far as a century before Peirce's birth. And, in my view, Peirce's importance resides not in the creation of this doctrine but in his transformation of it in subtle but significant ways. As I shall argue below, Peirce is the crucial, logical and historical link between nineteenth- and twentieth-century discussions of self-correction and progress towards the truth. Moreover, he is responsible for effecting a major metamorphosis in the self-correcting doctrine as it had been understood by his predecessors. To get some sense of the magnitude of that mutation, we must go back to the middle of the eighteenth century to see how and why the idea of self-correcting modes of inference arose.

Beginning in the 1730s and 1740s, a number of philosophers and scientists began to claim that science, as a result of the methods it employs, is a self-correcting enterprise. (Hereafter I shall refer to this view as the self-corrective thesis or simply SCT.)

Most early versions of SCT—like their more recent counterparts—were closely connected with a theory of scientific progress (SCT asserting, in effect, that science does "progress") and it is, therefore, not surprising that the Enlightenment view of intellectual progress first provided a leitmotif and rationale for SCT. That eclectic theory of knowledge, unique to the *philosophes,* which identified the growth of the mind with moral improvement of mankind, was certainly related to the doctrine of self-correction. But it is important not to be too beguiled by facile historical plausibilities. That the Enlightenment theory of progress produced fertile ground for the growth of the self-corrective view is quite likely; but we must look beyond the ethos of the Enlightenment to find the initial stimulus for theories of self-correction. Specifically, we must look to certain tensions and problems latent in the history of methodology itself. For instance, it is crucial to realize that the self-corrective thesis was itself a weakened form of a still more sweeping thesis which had dominated metascientific thought from antiquity. According to this more general thesis, which we might call the *thesis of instant, certain truth* (*TICT*), science—in so far as it is genuine science—utilizes a method of investigation which infallibly produces true theories. Virtually every theorist of method in the seventeenth century (including Bacon, Boyle, Descartes, Locke and Newton) subscribed to TICT.[4] The proponents of TICT believed that science could dispense with conjectures and hypotheses since there was, ready at hand, an "engine of discovery" (as Hooke called it[5]) which could infallibly (and usually mechanically) produce true theories. The concept of progress, within the framework of TICT, was clear and unambiguous. Progress, on this view, could only consist in *the accumulation of new truths*. The replacement of one partial truth by another simply made no sense in this context. Growth, in so far as it occurred, was by accretion rather than by attrition and modification.

By the middle of the eighteenth century, however, many methodologists were convinced that TICT was untenable. Difficulties in articulating a coherent logic of discovery, along with skeptical arguments about the inability of empirical evidence to prove a theory conclusively, conspired to chasten the scientist's confidence in the undisputed truth of his mental creations and to make (merely probable) hypotheses respectable, for the first time since the euphoria of the Scientific Revolution had made them unfashionable.[6]

There were two major arguments which seemed to undermine TICT: one was directed against the method of "proof *a posteriori*" (as Descartes had called it); the other, against eliminative induction.

The main argument of the first kind was an application of the so-called "fallacy of affirming the consequent" to scientific inference. As surprising as it might seem, several methodologists and scientists in the seventeenth and eighteenth centuries had argued that the ability of a theory to predict successfully an experimental result was *prima facie* evidence that the theory was a proven truth. Cartesians (e.g., Jacques Rohault) and Newtonians (e.g., Bryan Robinson) alike often slipped into this sloppy mode of reasoning. By 1750, however, the inconclusive character of this form of inference had been pointed out by Leibniz, Condillac, and David Hartley, among others.

Similarly, the method of proof by eliminative induction (associated with Bacon and Hooke) had been discredited by the arguments of Condillac, Newton, and LeSage against the possibility of exhaustively enumerating all the conceivable hypotheses which might be invoked to explain a class of events. These three all asserted that (in light of the impossibility of knowing that we have thought of all the appropriate hypotheses which might explain facts in a given domain) we can never be sure that the hypotheses which have survived systematic attempts at refutation are true.

(The third major candidate for a model of scientific inference, *enumerative* induction, had long since been discredited; in antiquity by Aristotle and Sextus Empiricus, and in early modern times by Bacon, Newton and Hume, among others.)

Since none of the known modes of "empirical inference" were valid, methodologists of science in the late eighteenth century were no longer able to speak, with a clear conscience, about the certainty and truth of scientific theories. (The notable exceptions to this generalization are the "a priorists" (e.g., Lambert, Wolff, and Kant); but their neo-Cartesian a priorism was a minority viewpoint.)

Prepared to concede that the theories of the day might eventually be refuted, convinced moreover that TICT was too ambitious, several late eighteenth-century methodologists produced a compromise. If, they reasoned, there is no instant, immediate truth, we can at least hope to reach truth *in the long run*. Even if the scientist's method does not guarantee that he can get the truth on the first attempt, perhaps he can at least hope to get ever closer to it. Even if the methods of science are not foolproof, perhaps they are at least capable of correcting any errors the scientist may fall prey to. Thus was born SCT. In some ways, it was a face-saving ploy, for it permitted the scientist to imagine that his ultimate goal was, as TICT had suggested, the Truth; although the scientist now had to be satisfied with the quest for ever closer approximations rather than the truth itself.[7]

At the same time that SCT was emerging (and this was no coincidence) some methodologists were moving away from a Baconian inductive model of scientific inference towards something like a model of conjecture and refutations.[8] Science was seen, not as a discipline where theories were somehow extracted or deduced from experiment, but rather as one where theories were formulated, tested, rejected and replaced by other theories. (I shall abbreviate this crude model as H-R-H′: hypothesis-refutation-hypothesis). When SCT was stated within the context of such a model of scientific inquiry, it generally amounted to the following claims:

(1) Scientific method is such that, in the long run, its use will refute a theory T, if T is false;

(2) Science possesses a method for finding an alternative T′ which is closer to the truth than a refuted T.[9]

On this view, which is as much an historiography as a philosophy of science, the temporal sequence of theories, T_1, T_2, . . . T_k, in any genuinely scientific domain is a series of ever-closer approximations to the truth (provided, of course, that science uses the method(s) which insure(s) self-correction). And there was a certain amount of intuitive plausibility to this picture. Even today, it is common to hear that Ptolemy's system was "closer to the truth" than Aristotle's system of concentric spheres; that Copernicus's heliocentric system was "more nearly true" than Ptolemy's; and that Kepler's elliptical system is a still closer approximation.

It is important to be clear about the problem or set of problems which SCT was presumed to resolve. Like TICT before it, *SCT was designed to provide an epistemic solution to the problem of scientific knowledge.* That problem can be put in various forms: Why should we take science seriously as a cognitive pursuit? What justification is there for the methods which science employs? Why should we prefer science to quackery or pseudo-science? Whatever our views about SCT, we must at least concede that it was an attempt to resolve what are perhaps the central problems of the philosophy of science; namely, the justification of both the knowledge-claims and the methods of the natural sciences.

Adherents of SCT provided what was, in its time, a highly original approach to this perennial problem. For them, the justification of science as a cognitive, intellectual pursuit was sought—not in the certainty or even the truth of its conclusions—but in its progressive evolution towards the truth. As I shall claim below, in the course of the later evolution of SCT, there was an increasing tendency to lose sight of this justificational problem in its full generality, a tendency to see the self-corrective thesis as the solution to rather different problems, of far less significance. But more of that below.

If the conditions I have spelled out indicate roughly what SCT amounts to, what was its rationale? What reason had Enlightenment philosophers to believe that science uses methods which satisfy conditions (1) and (2) above? The early proponents of SCT provided an answer, but not a very satisfactory one. Pursuing analogies between methods of mathematical inference and the methods of science, they claimed that just as the mathematician finds the roots of an equation by posing incorrect guesses, and then refining those guesses with mechanical tests, so the scientist can formulate an incorrect hypothesis and subsequently improve on it by comparing its results with observation, altering the hypothesis where necessary to bring it into closer agreement with fact. Clearly, the analogy is incomplete. After all, it does not prove that the methods of science are self-corrective to compare them with the admittedly self-corrective mathematical techniques unless the analogies between the two cases are very strong in appropriate respects. Unfortunately, they are not (or, at least, they were not shown to be) strongly analogous. Although it is relatively easy to show that the method of hypothesis-refutation-new hypothesis (H-R-H′) satisfies condition (1) above, there is no machinery for insuring that such a method satisfies condition (2) or even (2′).[10] Indeed, no methodological procedure was ever suggested in this period for replacing a refuted hypothesis by one which could be known to be closer to the truth. So impressed were these methodologists by the approximate techniques of mathematics that they did not worry about what (in our view) are the vast logical differences between scientific testing and mathematical proof. This perhaps can be made clear by discussing a pair of representative early defenders of SCT.

Among the first philosophers[11] to address themselves to this problem were David Hartley (1705–1757) and Georges LeSage (1724–1803), who, although working independently, arrived at almost identical results. I shall consider them in turn.

In a chapter, "Of Propositions, and the Nature of Assent," in his *Observations on Man* (1749), Hartley analyzes the sorts of methods which the scientist has at his disposal. Hartley insists that only in mathematics can one develop theories which can be rigorously demonstrated.[12] In science, however, we must be content with something less than immediate certainty. However, taking his cue from the mathematicians, Hartley believes that the scientist can utilize certain methods which, if they do not yield the truth immediately, will gradually bring the scientist to a true theory in the long run. He proposes two different methods, both based on mathematical techniques, both of which are self-correcting, and both of which are, in the long run, likely to lead the scientist to the truth.

1) *The rule of false position.* This approximate technique, known as the *regula falsa* among Renaissance mathematicians, is characterized by Hartley as follows:

Just as the arithmetician supposes a certain number to be that which is sought for; treats it as if it was that; and finding the deficiency or overplus in the conclusion, rectifies the error of his first position by a proportional addition or subtraction, and thus solves the problem; so it is useful in all kinds of inquiries, to try all such suppositions as occur with any appearance of probability, to endeavour to deduce the real phenomena from them; and if they do not answer in some tolerable measure, to reject them at once; or if they do, to add, expunge, correct, and improve, till we have brought the hypothesis as near as we can to an agreement with nature. After this it must be left to be further corrected and improved, or entirely disproved. . . .[13]

Two centuries earlier, the mathematician Robert Recorde had also been, like Hartley, impressed and amazed at the capacity of the rule of false position to generate truth from error, as this delightful piece of doggerel verse indicates:

> Gesse at this woorke as happe doth leade.
> By chaunce to truthe you may procede
> And first woorke by the question,
> Although no truthe therein be don.
> Such falsehode is so good a grounde,
> That truthe by it will soone be founde.[14]

Hartley takes Recorde's point one important step farther, however, by arguing that this sort of method works in natural philosophy as well as in algebra.

2) *The method of approximating to the roots of an equation.*
Like the rule of false position, Hartley sees this Newtonian technique as a means of generating a theory "which though not accurate,
approaches, however, to the truth."[15] Here, the scientist begins by a
guess at the root of the equation (viz., the correct answer). From
such a guess, applied to the equation, "a second position is deduced,
which approaches nearer to the truth than the first; from the second,
a third, etc."[16] Hartley insists that the use of such self-corrective
methods "is indeed the way, in which all advances in science are
carried on."[17]

There are, I believe, two important points to note about each of
these methods. In the first place, both involve the inquirer in making
posits (viz., hypotheses) which, if false, can be eventually falsified.
Much more importantly, they both provide a method, having once
refuted an hypothesis, for mechanically finding a replacement for it
which is closer to the truth than the original hypothesis. These two
characteristics together constitute the necessary and sufficient conditions for what I shall call a *strong self-correcting method (or
SSCM).*[18] A method is an SSCM if and only if (a) it specifies a
procedure for refuting a suitable hypothesis, and (b) it specifies a
technique for replacing the refuted hypothesis by another which is
closer to the truth than the refuted hypothesis.[19] Much of this paper
will be concerned with post-Hartleyan accounts of SCMs.

Unfortunately, Hartley himself does not indicate how we can
apply such mathematical methods to the natural sciences. While it
is easy enough to imagine that scientific hypotheses are refutable
(neglecting Duhemian considerations), it is more difficult to guess
what rule he had in mind for replacing a refuted scientific hypothesis
by one which was more nearly true.[20] Hartley simply takes it for
granted that one can, in a more or less straightforward fashion, import these mathematical techniques into the logic of the natural
sciences.

Hartley's contemporary, Georges LeSage, though drawing on
slightly different mathematical analogies, makes an argument very
similar to Hartley's. LeSage compares the procedure of the scientist
to that of a clerk solving a long division exercise. At each stage in
the division, we produce in the quotient a number which is more

accurate than the number appearing as the quotient in the preceding stage. At each stage, we multiply the divisor by the assumed quotient and see if it corresponds to the dividend. If it does not, we know that there is an error in the quotient, and we have a mechanical process for correcting the error, i.e., for replacing the erroneous quotient by one which is closer to the true value.[21] Going beyond such fanciful examples, LeSage, like Hartley, suggests that there are other approximative techniques which the scientist can borrow from the mathematician, including "the extraction of roots, the search for the rational divisors of an equation and several other arithmetical operations."[22] Beyond this, LeSage's views are, even to their ambiguity, sufficiently similar to Hartley's not to require separate consideration.

As I hinted before, the thesis that science is self-corrective and thereby progressive lends itself neatly to the eighteenth-century view of progress, for the sequence of theories of ever greater verisimilitude was the mirror image on the intellectual level of man's progressive perfection on the moral level. Joseph Priestly, who was in these matters a self-avowed disciple of Hartley, makes explicit the link between the self-corrective character of science and his theory of scientific progress. He writes:

Hypotheses, while they are considered merely as such, lead persons to try a variety of experiments, in order to ascertain them. These new facts serve to correct the hypothesis which gave occasion to them. The theory, thus corrected, serves to discover more new facts, which, as before, bring the theory still nearer to the truth. *In this progressive state, or method of approximation, things continue. . . .*[23]

Clearly, the weakness with all these programmatic statements is that they simply insist that scientific methods *are* self-corrective, without indicating precisely the manner in which they are so. Without a persuasive reason for believing that the methods of science are self-corrective, we have no rational grounds for speaking of scientific progress, a point which the logician and physiologist Jean Senebier was quick to emphasize: "Often we move imperceptibly away from the truth, and do so even whilst we believe that we are working towards it."[24] The case against the vagueness of SCT as developed by Hartley and LeSage was put convincingly by Pierre Prevost in 1805. He insists that scientific procedures necessarily differ from such mathematical techniques as the rule of false position. He observes that we do not generally have the knowledge in science to be able to satisfy the conditions of the rule of false position, and that we therefore cannot expect much from that method in science. Prevost argues specifically against the self-correcting character of the H-R-H′ method. All that method permits us to do, in his view, is verify or refute an hypothesis; it provides no machinery for replacing a refuted hypothesis with a better one:

Thus when Kepler, beginning with the circular hypothesis, tried out various eccentricities for the orbit of Mars, these false suppositions could (and indeed should) never have led him to a solution. When afterwards he recognized the weakness in the circular hypothesis, if he had tried other curves entirely by chance, he would have been using another method which could well have never brought him to his goal.[25]

I hope these few texts have made it reasonably clear that by the early years of the nineteenth century, the problem of justifying scientific knowledge (i.e., as infallible, indubitable truth) had been replaced—at least among some writers—by a program for justifying science by claiming that it pursues a method which will lead it ever closer to the truth. The extent to which this kind of approach came to dominate methodological thought is illustrated by the fact that the philosophies of science of Herschel, Comte, and Whewell are all concerned overtly with the progress of science and its gradual approach to the truth.[26]

Among nineteenth-century scientists as well as methodologists, the view persisted of science as an enterprise moving inexorably closer to a final truth. Claude Bernard, among others, conceived science in this approximative way. Thus, Ernest Renan wrote about Bernard:

Truth was his religion: he never had any disillusionment or weakness, for not a moment did he doubt science. . . . The results of modern science are not less valuable for being acquired by successive oscillations. These delicate approximations, this successive refining, which leads us to modes of understanding *ever closer to the truth* are [for Bernard] the very condition of the human mind.[27]

Similarly, that fervent Darwinian T. H. Huxley believed that "the historical progress of every science depends on the criticism of hypotheses—on the gradual stripping off, that is, of their untrue or superfluous parts. . . ."[28]

The key to the progressiveness of science was thought to reside in the fact that it utilized a method which was essentially self-corrective in character. Given time and sufficient experience, science could be perfected to any stage desired. In the middle years of the nineteenth century, especially with Comte and Whewell, the doctrine of progress through self-correction became in many ways the central concern of the philosophy of science. Science was seen as a growing, dynamic enterprise and, accordingly, philosophers of science were prone to stress such dynamic, growth-oriented parameters as increasing scope and generality, greater accuracy and systematicity and, above all, progress towards the truth. However, throughout much of the nineteenth century, the self-corrective character of scientific method, while regularly invoked and persistently praised, remains as unestablished as it had been with LeSage and Hartley. Everyone assumes that science is self-corrective (and thereby progressive), but no one bothers to show that any of the methods actually being proposed by methodologists are, in fact, self-corrective methods.

The focus of the self-correcting thesis had always been on conceptual change, on the progressive succession of one theory by another. What self-correctionists had sometimes ignored was that sort of "progress" which comes from increasing the probability of theories (most often by successful confirmations), without any change in the theories themselves. This second type of progress, which we might call "progress by probabilification," received much attention in the nineteenth century. Herschel, Brown, Whewell, Jevons and Apelt (to name only a few) discussed at length the methods by which we can gain confidence in our theories, without necessarily altering them. Partisans of progress through probabilification tended to stress the continuity of scientific theory; for them, experiments with high confirming potential were emphasized rather than the falsifying experiments which the self-correctionists stressed. If the advice of the self-correctionists to experimental scientists was "Devise experiments which will indicate weaknesses in your theories," the corresponding advice from the probabilifications was "Devise experiments which, if their outcome is favorable, will do most to contribute to the likelihood of your theories." Impressed by Laplace's rule of succession and the application of probability theory to induction, the "probabilists" argued that every valid theory goes through all the degrees of certainty from extreme improbability to great likelihood. (Writers like Thomas Brown and John Herschel identified that transition as one from "hypothesis" to "theory" or "law.")

It would be wrong to give the impression that these two alternative theories of scientific progress, one by self-correction and the other by probabilification, were mutually exclusive. On the contrary, several of the best methodologists of the period (e.g., Whewell and Bernard) adopted both, arguing that "local" progress occurred by probabilification, while "cross-theoretical" progress was governed by a self-corrective method.[29] These two approaches did, however, represent different emphases, and were to give rise in the twentieth century to two very different strains in philosophy of science (Carnap and Keynes being the descendants of the progress by probabilification school, and Popper and Reichenbach focusing primarily on progress by self-correction).

A third theory of scientific progress prominent in the nineteenth century was that endorsed by Mill and Bain. Mill adopts a theory of progress by elimination. An hypothesis is entertained, tested, refuted and replaced by another one. This perhaps seems but another version of the standard H-R-H'. But it receives an interpretation by Mill very different from that of the self-correctionists. Mill does not believe we have any good grounds for believing that the H' is any more true than the refuted H. Indeed, it may be "more false." But the sequence of hypotheses H, H', . . . H^n is a progressive one because the last remaining member of the series is true. Adhering to a principle of limited variety, Mill maintained that there were only a finite number of candidates for the status of a scientific law and the false contenders could be eliminated by a judicious use of the five canons of induction. Clearly, Mill's account of scientific progress differs substantially from that of both the self-correctionists and the probabilificationists.

All three of these theories of scientific progress found their followers in the second half of the nineteenth century. Nonetheless, the self-correctionists predominate, and it is late nineteenth-century developments in the self-corrective tradition which I want to examine now.

As we have seen, for more than a century after Hartley and Le-Sage, methodologists almost to a man (Mill being the most noteworthy exception) endorsed SCT and, ignoring the doubts voiced by Senebier and Prevost, assumed without much argument that the methods of science in general, and H-R-H' in particular, were genuinely self-corrective. The discussion of this question was given an entirely new slant, however, by the work of Charles Sanders Peirce, whose approach to this question I wish to discuss in some detail.

It is well known that Peirce was a persistent defender of SCT. Unlike most of his predecessors, however, Peirce (usually) realized that SCT was not self-evidently true, and felt that one of the tasks of the logician of science was to show how and why science is a self-corrective enterprise which, in its historical development, gradually but inexorably comes closer and closer to a valid objective representation of natural phenomena.[30]

Peirce's most crucial claim in this regard is his insistence that *all scientific inquiry is self-corrective in nature.* "This marvelous self-correcting property of Reason," he writes, "belongs to every sort of science . . . ,"[31] and every branch of scientific inquiry exhibits "the vital power of self-correction."[32] The reason the sciences are self-corrective is that (in Peirce's view) they utilize methods which are self-corrective. It is thus incumbent on Peirce to show that all the methods of science exhibit self-correction and thereby guarantee progress towards the truth. Those methods for Peirce are threefold: deduction, induction and abduction.

It is at this point that the first of Peirce's serious problem slides occurs. Although he is presumably obliged to show that all three methods of science are self-corrective, he ignores deduction and, less excusably, abduction, and limits his discussion almost entirely to induction. There is, nonetheless, a certain rationale for this since, in Peirce's view, inductive methods are operative in every appraisal we make of a theory. So long as the inductive step is self-corrective, any failure of self-correction in deduction and abduction may be ameliorated. Thus, Pierce's problem is changed from that of showing that scientific methods generally are SCMs, to demonstrating that the various methods of induction are self-corrective. The "induction" referred to in that slogan is not any one single technique, but rather the entire machinery for the testing of a scientific hypothesis. Although the precise significance of the term "induction" undergoes several notorious shifts in his long career, this very general sense of the term is a persistent feature of almost all his discussions of the question.

Thus, in about 1901, Peirce wrote that "the operation of testing a hypothesis by experiment . . . I call *induction*."[33] In 1903 he virtually repeated this definition,[34] and in his later, important essay on "The Varieties and Validity of Induction" (c. 1905) he makes substantially the same point:

The only sound procedure for induction, whose business consists in testing hypotheses . . . is to receive its suggestions from the hypothesis first, to take up the predictions of experience which it conditionally makes, and then try the experiment . . . when we get to the inductive stage what we are about is finding out how much like the truth our hypothesis is. . . .[35]

Peirce asserts on any number of occasions that induction conceived in this broad sense is self-corrective in nature. As early as 1883, he observed that: "We [must not] lose sight of the constant tendency of the inductive method to correct itself. This is of its essence, this is the marvel of it."[36] He reiterates this point twenty years later: "[Induction] is a method of reaching conclusions which, if it be persisted in long enough, will assuredly correct any error concerning future experience into which it may lead us."[37] Between these two temporal extremes, Peirce regularly returns to SCT. About 1896 for instance, he remarks that "Induction is that mode of reasoning which adopts a conclusion as approximative [approximately true], because it results from a method of inference which must generally lead to the truth in the long run."[38] And two years later he smugly claims that the fact "that induction tends to correct itself, is obvious enough."[39] To this point, the Peircean texts I have cited could have been written by LeSage, Hartley, Whewell or any of a dozen other methodologists living in the century before Peirce.

What Peirce perceived, which his predecessors had not, was that it was not all that obvious that induction, defined as the testing of an hypothesis, is, or tends to be, self-correcting. He saw this as a genuine problem and one which he attempted to resolve on several occasions, most notably in the Lowell Lectures of 1903, and the famous manuscript "G" (c. 1905). In his classic essay of 1903, Peirce distinguished three varieties of induction: *crude* induction, *qualitative* induction, and *quantitative* induction. *Crude* induction is concerned with universal (as opposed to statistical) hypotheses, the evidential base for which is flimsy and precarious in that they are merely empirical generalizations of the type "all swans are white" or "all Germans drink beer." What typifies crude inductions is not so much the logical form of their conclusions as the nature of the evidential base on which they rest. The only license required for making a crude induction of the form "All A are B" is "the *absence* of [any known] instances to the contrary."[40] Such inductions are indispensable to daily life but, on Peirce's view, they play no significant role in science. *Quantitative* induction, on the other hand, is an argument from the observed distribution of certain properties in a sample to an hypothesis about the relative distribution of those properties in a larger population. Quantitative induction is induction by simple enumeration in its most literal sense. The conclusion of a quantitative induction is always a statement concerning the probability "that an individual member of a certain experiential class, say the S's will have a certain character, say that of being P."[41] Unlike crude induction, quantitative induction is (according to Peirce) used in the sciences, if only to a limited extent.

"Of more general utility" is the remaining variety of induction, *qualitative* induction.[42] This corresponds, more or less, to what is usually called the hypothetico-deductive method. Here, the scientist formulates an hypothesis, deduces predictions from it, and performs experiments to check the predictions. If all of the tested predictions are confirmed, this hypothesis should be tentatively adopted; while if any of the predictions are refuted, the scientist modifies the hypothesis, or abandons it and tries another. Clearly, qualitative induction corresponds to the method I have been denoting by H-R-H'.

Peirce then proceeds to argue that one of these species of induction, namely the quantitative variant, is genuinely self-corrective.

"Quantitative induction," he insists, "always makes a gradual approach to the truth, though not a uniform approach."[43] Peirce's argument for the self-correcting character of quantitative induction is a crude version of the arguments advanced more recently by Reichenbach and Salmon. Provided that our sampling procedures are fair and that our long run is long enough, the estimates which quantitative inductions lead us to posit will in time approximate ever more closely to the true value.[44] (In developing this argument, Peirce tells us that he was impressed, as Hartley and LeSage had been 150 years earlier, by the fact that "certain methods of mathematical computation correct themselves.")[45]

Ignoring the familiar technical difficulties with this argument,[46] let us concede that Peirce came close to showing that quantitative inductions are self-corrective. At all events, quantitative induction does satisfy two conditions for a self-correcting method; namely, it is a method which not only allows for the refutation of an hypothesis but which also mechanically specifies a technique for finding a replacement for the refuted hypothesis[47] (provided, and it is a crucial proviso, that the hypothesis is taken as a probability sentence).

But what of that scientifically more significant species of induction, H-R-H'? Such qualitative inductions clearly satisfy the first condition for an SCM, insofar as persistent application of the method of hypothesis testing will eventually reveal that a false hypothesis is, in fact, false. But the method of qualitative inductions provides no machinery whatever for satisfying the second necessary condition for an SCM; *given that an hypothesis has been refuted, qualitative induction specifies no technique for discovering an alternative H' which is (or is likely to be) closer to the truth than the refuted H*. Nor does it even provide a criterion for determining whether an alternative H' is closer to the truth than H. Peirce, in short, gives no persuasive arguments to establish that qualitative induction is either strongly or weakly self-corrective.[48]

At a certain level of consciousness, Peirce was fully aware of the fact that he had not shown qualitative inductions to be self-corrective. He remarks that while quantitative induction "always makes a gradual approach to the truth . . . qualitative induction is not so elastic.

Usually either this kind of induction confirms the hypothesis or else the facts show that some alteration must be made in the hypothesis."[49] What the facts do not show, of course, is how the hypothesis is to be altered so as to bring it closer to the truth. While "the results of [qualitative] induction *may* help to suggest a better hypothesis," there is no better one.[50]

And in one especially candid lecture (1898) on the "Methods for Attaining Truth," Peirce confesses that in "the Explanatory Sciences," we have no sure way of knowing whether the outcome of any confrontation between competing theories is "logical or just."[51] Peirce has evidently landed himself in a situation in which he is pursuing a rapidly degenerating problem. Where before he had answered the question "Are the methods of science self-corrective?" by replying that at least all the inductive methods of science are self-corrective, he is here reduced to saying that even of the various methods of induction, only one is known to be genuinely self-corrective.

Peirce must have sensed the awkwardness of the position in which he found himself. Having set out to show that science is a progressive, self-corrective enterprise, moving ever closer to the truth—and there can be no doubt that this was his initial problem, since both the tradition he was in and his early writings make this clear—Peirce finds himself able to show only that one of the methods of science (and that, by Peirce's admission, a relatively insignificant one) was self-corrective.

I cannot stress too strongly how important it is to be clear about Peirce's intentions. Virtually all Peirce's recent commentators have seen him as setting out to answer Hume's doubts about induction; and have, accordingly, discussed his accounts of SCT and enumerative induction as if they were intended only or primarily as an answer to Hume. Unless my analysis is completely wrong-headed, this is to judge Peirce by an inappropriate yardstick. It was not enumerative induction, but science which Peirce set out to justify; it was not Hume but the cynical critics of science whom Peirce set out to answer. (I might generalize this point by adding, parenthetically, that it is one of the wilder travesties of our age that we have allowed the myth to develop that nineteenth-century philosophers of science were as preoccupied with Hume as we are. I will even wager that when the definitive history of nineteenth-century philosophy of science is written, Hume's account of induction will loom about as large therein as Bradley's account of metaphysics would in a comparable history of twentieth-century philosophy of science. As far as I have been able to determine, none of the classic figures of nineteenth-century methodology—neither Comte, Herschel, Whewell, Bernard, Mill, Jevons, nor Peirce—regarded Hume's arguments about induction as much more than the musings of an eclectic historian. (This claim is borne out by the fact that in Peirce's thirty-two papers on induction and scientific method—papers teeming with historical references—there is only one reference to Hume; and that is not in connection with the problem of induction but with the problem of miracles.)[52] I suspect that this point, if historically correct as I believe it is, has serious ramifications for an understanding of the whole Peircean enterprise.)

As it turned out, Peirce attempted to bridge the gap between intention and performance by a combination of bluster and repetition. Just as LeSage and Hartley could, a century earlier, gloss over their failure to demonstrate an analogy between approximative techniques in mathematics and the methods of science, so Peirce conveniently ignores his painstaking discrimination between the various forms of induction, and pretends (as the quotations above make clear) *that his argument has established that all forms of induction (and, by implication, all scientific inferences) are SCMs.* In his later writings,[53] he will generally assert that qualitative inductions (or, as he sometimes calls them, Inductions of the Second Order) are progressive and self-corrective; but he never goes further than asserting that such methods are SCMs, without even the pretense of an argument for that assertion.

Lenz has charitably said that Peirce's "remarks on the self-correcting nature of the broader form[s] of induction are extremely hard to comprehend."[54] I think we must lay a more serious charge at Peirce's feet than that of obscurity. Peirce's remarks in themselves are not difficult to comprehend; he says quite plainly that all forms of induction are self-corrective. What *is* hard to comprehend is Peirce's reason for making such a general assertion. And I think it would be less than candid not to say that Peirce offers no cogent reasons, not even mildly convincing ones, for believing that most inductive methods are self-corrective. I suspect that the explanation of this glaring oversight may be found by recalling Peirce's original motivation.[55]

Peirce began, as I claimed before, with a very general and a very interesting problem: that of justifying scientific inference by showing that the methods of science (including all species of induction) are self-corrective. This was, as I have shown, one of the standard problems of philosophy of science by Peirce's time. Unable to find a general solution to that problem, Peirce tackles the more limited task of showing that one family of inductive arguments, quantitative inductions, are self-corrective. Having shown (at least to his own satisfaction) that quantitative induction is self-corrective, Peirce then, without even the hint of a compelling argument, makes the crucially serious slide. Seemingly unwilling to admit, even to himself, that he has failed in his original intention to establish SCT for all the methods of science, Peirce acts as if his arguments about quantitative induction show all the other species of induction to be self-corrective as well.

His dilemma was genuine. Having discovered that he could only show quantitative induction to be self-corrective, he could have gone the way of Reichenbach and argued that quantitative induction was the only species of scientific inference, to which all other legitimate methods could be reduced. But Peirce did not share Reichenbach's belief that complex inference was a composite of simple inductions by enumeration. Alternatively, he could have abandoned SCT altogether, conceding that science uses methods which are not, so far as we know, self-corrective. But that would have meant taking much of the flesh out of his philosophy of science. Faced with two such debilitating alternatives, Peirce conveniently ignored the restricted scope of his argument and (perhaps unconsciously) slid from the self-corrective character to the straight rule to SCT as a general thesis. The extent to which Peirce was prepared to make this leap is illustrated by such remarks as his claim that "inquiry of every type, fully carried out, has the vital power of self-correction and growth."[56]

At one point, his bedrock commitment to SCT, even in the absence of any methodological rationale for it, becomes clear:

It is certain that the only hope of retroductive reasoning [viz., qualitative induction] ever reaching the truth is that there may be some natural tendency toward an agreement between the ideas which suggest themselves to the human mind and those which are concerned in the laws of nature.[57]

Unable to find a rational justification for his intuition that science is self-corrective, the otherwise tough-minded Peirce had to fall back on Galileo's *il lume naturale,* on an inarticulate faith in the ability of the mind somehow to ferret out the truth, or a reasonable facsimile thereof:

We shall do better to abandon the whole attempt to learn the truth . . . unless we can *trust* to the human mind's having such a power of guessing right that before many hypotheses shall have been tried, intelligent guessing may be expected to lead us to the one which will support all test. . . .[58]

Such a belief was shared by Peirce's contemporary Pierre Duhem, who argued for SCT in terms of an approach to "*the* natural classification." In a more explicit manner than Peirce, Duhem concedes that he can produce no logically compelling grounds for believing that the history of science brings us closer and closer to a genuine representation of natural relations. Nonetheless he is convinced that this occurs and that every scientist knows that SCT is true:

Thus, physical theory never gives us the explanation of experimental laws . . . but the more complete it becomes the more we apprehend that the logical order in which theory orders experimental laws is the reflection of an ontological order, the more we suspect that the relations it established among the data of observation correspond to real relations among things. . . . The physicist cannot take account of this conviction. . . . But while the physicist is powerless to rid his reason of it . . . yielding to an intuition which Pascal would have recognized as one of those reasons of the heart "that reason does not know," he asserts his faith in a real order reflected in his theories more clearly and more faithfully as times goes on.[59]

Less optimistic than Peirce about the possibility of finding a methodological rationale for the view that science moves ever closer to the truth, Duhem maintains that the methodologist cannot justify SCT, and that its only defense lay in what Duhem calls a "metaphysical assertion."[60]

To return to Peirce only briefly, I suspect that there is another important sense in which he takes much of the force out of the SCT tradition. As I have tried to make clear, that tradition was committed to the view (among others) that the replacement of one non-statistical hypothesis by another was the basic unit of progress and self-correction. Peirce, at least on some occasions, abandons that view altogether. In its place, he argues that, although we have no way of correcting our hypotheses, what we can correct are the assignments of probability which we give to those hypotheses. When arguing in this vein, Peirce sees the process of assigning probabilities to hypotheses as self-corrective, while the process of replacing one hypothesis by another no longer remains even a candidate for consideration as a self-corrective process. In the course of time, it is not our theories which get closer to the truth, but rather, the probabilities which we assign to theories exhibit progress and self-correction. Where all previous discussions of the question had been concerned to show that a sequence of hypotheses of the form:

$$A \text{ is } B,$$
$$A \text{ is } C,$$
$$A \text{ is } D,$$
$$\text{etc.}$$

is progressive and self-corrective, Peirce's quantitative induction goes for the "cheapest" form of self-correction, arguing that a sequence of the following kind:

$$\text{The probability that } A \text{ is } B \text{ is } m/n,$$
$$\text{The probability that } A \text{ is } B \text{ is } m'/n',$$
$$\text{The probability that } A \text{ is } B \text{ is } m''/n'',$$
$$\text{etc.}$$

is (or can be) self-corrective. Peirce simply cannot handle a case where an hypothesis (of the form "A is B") is replaced by a conceptually different one (say, "A is C").[61]

II

It would not be appropriate in this volume to discuss at length the views of more recent methodologists about SCT, since that would take us well into the twentieth century. Nonetheless, I think a few words are in order about more recent developments in so far as they link up rather closely to the tradition I have been discussing here. As everyone knows, Hans Reichenbach took up SCT, most notably in his *Wahrscheinlichkeitslehre* and his *Experience and Prediction*. In both works, Reichenbach, like Peirce, sets out to show that the straight rule is self-corrective, that induction by simple enumeration is an SCM. Like Peirce, Reichenbach then goes on to assume, with only the flimsiest of arguments, that science is a self-correcting enterprise because (and here Reichenbach differs from Peirce) all the methods of science can be reduced to enumerative induction.[62] Unfortunately, however, Reichenbach's attempts to reduce most scientific methods to convoluted species of the straight rule are at best, programmatic; at worst, unconvincing. As a result Reichenbach, like Peirce, found himself unable to prove SCT generally, and was forced to be content with the comparatively insignificant consolation that enumerative induction is self-corrective.

All the same, it must be said on Reichenbach's behalf that he takes up the banner of the SCT tradition in a less half-hearted way than Peirce had. Reichenbach sensed the object of the exercise, and understood that exploration of the self-correcting properties of the straight rule was only of crucial import in so far as one could establish the relevance of the straight rule to more subtle forms of scientific reasoning. That Reichenbach's program did not come off, that he never quite managed to achieve the reduction of scientific methods to enumerative induction, does not diminish the soundness of his intuitions as to the nature of the problem.

In the last two decades, however, there seems to have been a tendency to return to a Peircean rather than a Reichenbachian treatment of the question. Many contemporary philosophers of science, perhaps forgetting that self-correction was originally a thesis about science rather than a putative answer to Hume, have explored at length the question whether the method of enumerative induction is self-corrective without asking the wider and crucial question whether science exemplifies inductions, without seriously considering whether the methods of science are enumerative.[63] Reichenbach's most distinguished disciple, Wesley Salmon, similarly skirts this particular issue on many occasions.[64] One has the impression (perhaps unjustifiably) that such philosophers have become so involved with the technical and formal aspects of Peirce's solution that they have lost sight of the problem to which it was a solution. We are, I suspect, sometimes repeating Peirce's mistake of thinking that so long as we establish that any ampliative inference is self-corrective, we can easily show that most of them are.

Criticisms of the type I have offered here, however well intentioned, are always open to the charge of being premature and philistinistic. After all, one might say, a break-through could come at any moment and in that event the work of Reichenbach and Salmon might become to the foundations of scientific inference what Russell, Frege or Cantor were to the foundations of mathematics. Moreover, it might be pointed out that foundational studies, especially in their preliminary stages, always have only tenuous connections to what they purport to be foundations of. But, granting all that, one has a right to insist that putative "foundational studies" must satisfy some canons of adequacy, and be subject to certain standards of criticism.

Precisely what these standards are, I do not pretend to know. (This in itself is a major philosophical problem.) But there are several seemingly relevant points to make about the so-called pragmatic justification of induction and scientific inference. The first point is that distinguished philosophers have been exploring this approach for almost a century. In that time, they are no closer to exhibiting a connection between the straight rule and other modes of inference than Peirce was in 1872. While promissory notes are not dated, there is at least a presumption that payments will be made at respectable intervals. Secondly, and more disturbingly, the pragmatic approach has, at least since the 1930s, tended to concern itself less and less with the one thesis which originally made that approach interesting, viz., the thesis that scientific inference could be reduced to enumerative inference. The centrality of that thesis in the pragmatic tradition has been replaced by an exaggerated preoccupation with enumerative induction. In an unnoticed sleight of hand, the problem of the justification of science has been displaced by the problem of justifying induction. And, in the absence of any established link between the former and the latter, this portion of the "philosophy of science" has surrendered any convincing claims to being the philosophy of science.

If we believe, with Peirce, LeSage, Hartley, Whewell, and Duhem that science is a self-corrective, progressive enterprise, then we should presumably be seeking to show how and why it is so. If we further believe with Peirce and Reichenbach that the exploration of enumerative induction will provide the answer, then we ought to be exploring more assiduously the role of enumerative induction in actual science. What we must avoid is falling into the Peircean pit by assuming without argument that the grand old problem of the progress of science will necessarily (or even probably) be clarified by technical investigations of the straight rule. We have accepted Peirce's *ersatz* self-correction—a self-correction which can only talk about changes in probabilities rather than changes in theories—without openly discussing whether full-bodied self-correction in the traditional sense is beyond our powers of explication. It is the self-correcting nature of science, not the self-corrective nature of a "puerile" rule, which should be our main concern.[65]

NOTES

1. *Mid-West Quarterly* 2 (1914): 49.
2. There is a vast body of exegetical and critical literature dealing with Peirce's philosophy. The following are concerned explicitly with Peirce's treatment of self-correction:

A. W. Burks, "Peirce's Theory of Abduction," *Philosophy of Science* 13 (1946): 301–306.
C. Y. Cheng, *Peirce's and Lewis's Theories of Induction* (The Hague, 1969).
H. G. Frankfurt, "Peirce's Notion of Abduction," *Journal of Philosophy* 55 (1958): 593–597.
J. Lenz, "Induction as Self-Corrective," in Moore and Robin, eds.,

Studies in the Philosophy of Charles Sanders Peirce (Amherst, Mass., 1964).

E. Madden, "Peirce on Probability," in ibid.

F. E. Reilly, *The Method of the Sciences According to C. S. Peirce.* Doctoral dissertation, St. Louis University, 1959.

While acknowledging a significant debt to all of these works, I believe it is fair to say that none of these authors treats Peirce's approach to SCT within the historical framework in which I have tried to place it.

3. See, for instance, Burks, "Peirce's Theory." Even Peirce himself tries to give the impression that he was the first to enunciate the view that scientific reasoning is self-corrective. For instance, he wrote in 1893 that "you will search in vain for any mention in any book I can think of" of the view "that reasoning tends to correct itself." C. S. Peirce, *Collected Papers,* eds. Hartshorne, Weiss et al., 8 vols. (Cambridge: Harvard University Press, 1931–58) 5:579. Without questioning Peirce's integrity, we do have some grounds for doubting his memory. Peirce makes numerous references to the works of many of the writers whom I cite below as Peirce's predecessors in this matter. (See, for example, ibid., 5:276n, where he writes knowledgeably of the philosophies of science of both LeSage and Hartley, who had stressed the self-correcting aspects of scientific reasoning.)

4. This point requires some qualification. As is well known, passages can be adduced from all these authors where they seem to abandon the infallibilism of TICT and to replace it by a more modest "probabilism." (Many of the relevant texts are discussed in my "The Clock Metaphor and Probabilism," *Annals of Science* 22 [1966]: 73–104.) However, it would be a serious error of judgment to let these concessions to fallibilism obscure the fact that all of these figures shared the classical view that science at its best is *demonstrated knowledge from true principles.* Bacon, Descartes, Locke and Boyle all see it as a goal that science become infallible; until that goal is realized they are willing to settle—but only temporarily—for merely probable belief. Their long-term aim, however, is to replace such mere opinion by genuine knowledge.

5. See Robert Hooke's posthumously published account of "inductive logic" in *The Posthumous Works of Robert Hooke,* ed. R. Waller (London, 1705), pp. 3ff.

6. See my "The Clock Metaphor and Probabilism," *Annals of Science* 22 (1966): 73–104.

7. A century and a half later Max Planck gave eloquent expression to this quintessentially eighteenth-century viewpoint: "Nicht der Besitz der Wahrheit, sondern das erfolgreiche Suchen nach ihr befruchtet und beglüchte den Forscher." (*Wege zur physikakischen Erkenntnis,* 4th ed. [Leipzig, 1944], p. 208.)

8. Some, but by no means all. As late as the 1790s, philosophers such as Thomas Reid were still arguing for a strictly inductive methodology. (Cf. my "Thomas Reid and the Newtonian Turn of British Methodological Thought," in *The Methodological Heritage of Newton,* ed. Butts and Davis [Toronto, 1969].)

9. To be faithful to the historical situation, it is important to point out that some eighteenth- and early nineteenth-century methodologists, while accepting SCT as a general thesis, were not altogether happy with the idea expressed in (2) above. As formulated there, SCT is committed to the view that there is a *mechanical* process for finding alternatives. Some methodologists denied this. What they did insist on, however, was that (2'): Science possesses techniques for determining unambiguously whether an alternative T' is closer to the truth than a refuted T.

William Whewell, for instance, denied the claim implicit in (2) that the scientist possessed any algorithm for automatically correcting an hypothesis. Nonetheless, he was convinced that it was generally possible, given a (refuted) theory and an alternative to it, to determine which of the two was (in Whewell's language) "nearer to the truth." Hereafter, I shall refer to the pair (1) and (2) as the *strong* self-correction thesis (or SSCT) and to the pair (1) and (2') as the *weak* thesis of self-correction (WSCT).

There is another important qualification to make here. Although all the figures I discuss talk about "getting closer to the truth," "moving nearer to the truth," etc., it is not altogether clear that there is a shared conception of what truth consists in. With some writers, for instance, the notion of truth seems to be an instrumental one (viz., that is true which adequately "saves the phenomena"); with others, the concept of truth is closer to the Aristotle-Tarski line. Nonetheless, most discussions of self-correction and proximity to the truth seem to be conducted independently of various conceptions of, and criteria for, the truth.

10. See Note 9 above.

11. This claim for the priority of LeSage and Hartley is, like all claims for historical priority, necessarily tentative.

R. V. Sampson, in his *Progress in the Age of Reason* (London, 1956), asserts that Blaise Pascal conceived of science as "cumulative, self-corrective" and progressive. I have been unable to find such an argument in Pascal and (unfortunately) Sampson offers no evidence for his interpretation. Similarly, Charles Frankel (*The Faith of Reason,* [New York, 1948]) argues, likewise without evidence, that "For Pascal . . . scientific method was progressive because it was public, cumulative, and self-corrective" (p. 35). Until more substantive evidence is produced, I believe the available historical evidence supports my priority claims for Hartley and LeSage. However, the argument in the body of the essay does not depend on the priority issue.

12. David Hartley, *Observations on Man* (London, 1749), 1:341–2.

13. Ibid., 1:345–6. Basically, the rule of false position worked as follows: If one sought the solution to an equation of the form $ax + b = 0$, one made a conjecture, m, as to the value of x. The result, n, of substituting m for x in the left-hand side of the equation is given by $am + b$. The correct value of x was then determined by the formula

$$x = mb/(b - n)$$

The rule of false position was one of the earliest known rules for the solution of simple equations.

It should be added that during the eighteenth century, the term "rule of false position" normally referred, not specifically to the rule given above, but rather to what we call the rule of double position, which involves two conjectures rather than one. An interesting discussion of this latter rule may be found in Robert Hooke's *Philosophical Experiments and Observations,* ed. Durham (London, 1726), pp. 84–86.

14. R. Recorde, *Ground of Artes* (London, 1558), fol. Z4.

15. *Observations on Man,* n12, 1:349.

16. Ibid., 1:349. It is perhaps worth observing that Hartley adhered to a theory of moral progress and self-improvement which paralleled the progress and self-correction of science. "We have," he writes, "a Power of suiting our Frame of Mind to our Circumstances, of correcting what is amiss, and improving what is right" (*Observations on Man,* 1:463).

17. Ibid., 1:349. There is, we should observe, a very great difference in the results which these various "approximative methods" yield. Some of these methods—such as the Newtonian method of approximation to the roots of a general equation—do not necessarily ever yield a true result. We can, by their use, constantly improve our estimate, but there is no guarantee that we will ever determine precisely the correct answer. However, other methods Hartley mentions, especially the rule of false position, not only correct a false guess, but immediately replace it by the true solution. These differences become very significant when applied to a scientific context. If our model for scientific method is the rule of false position, then one can imagine science rapidly reaching a stage where all the false theories have been replaced by true ones, and where scientific knowledge would be both static and non-conjectural. If, on the other hand, our model for inquiry is the search by approximation for the roots of an equation, then science would seem to be perhaps perennially in a state of change and flux, with no guarantee whatever that it could ever reach the final truth.

Hartley, as well as most of his nineteenth-century successors, seems to vacillate between these two very different models.

18. Talk of a "self-correcting method" is, of course, slightly misleading since the method does not correct itself, but rather it allegedly

corrects those statements which an earlier application of the method produced. However, since linguistic traditions sanctify all manner of confusions, and since it is *de rigueur* to speak of methods with these properties as self-corrective methods, I will do so, hoping the reader will bear this *caveat* in mind.

19. A method will be weakly self-corrective (WSCM) if (a) above and if (b) without itself specifying a "truer" alternative, it can determine for certain whether a given alternative is "truer." (See also Note 6 above.)

20. Precisely this criticism was raised by Condillac in 1749 against the view that science can borrow the approximative methods of the mathematician. (Cf. his *Traité des Systèmes* [Paris, 1749], pp. 329–31.) It was also raised by J. Senebier a generation later. (Cf. his *Essai sur l'Art d'Observer et de Faire des Expériences* [Geneva, 1802], 2:215–216.)

21. As LeSage puts it:

The corrections made of these particular suppositions, resulting from the small multiplications which serve to test their validity, have as their sole aim to bring closer together these suppositions and the [true] number; with the exception of the last partial division, which must be performed rigorously because it is here that one finally rejects the inaccuracies one has permitted oneself in the previous operations.

G. H. LeSage, "Quelques Opuscules rélatifs à la Méthode," posthumously published by Pierre Prevost in his *Essais de Philosophie* (Paris, 1804), 2:253–335. The passage in question dates from the 1750s, and appears on p. 261. (I discuss LeSage's work at much greater length in my "Georges LeSage: A Case Study in the Interaction of Physics and Philosophy," *Proceedings of the Second International Leibniz Congress* [Hanover, forthcoming], vol. 4.)

22. Ibid., 2:261.

23. Joseph Priestley, *The History and Present State of Electricity* (London, 1767), p. 381.

24. "Souvent on s'écarte du vrai, sans douter, et on le fuit en croyant le poursuivre." (*Essai,* 2:220.)

25. Prevost, *Essais de Philosophie* (Paris, 1804), 2:196. Prevost nevertheless believes that there are self-corrective methods which the scientist can use.

26. See, for instance, the several essays on progress in Whewell's *Philosophy of Discovery* (London, 1860) and Auguste Comte's preliminary discourse to the *System of Positive Polity* (4 vols. [London, 1875–7]). Similar, if more vague, sentiments are involved in John Herschel's discussion (*Preliminary Discourse on the Study of Natural Philosophy* [London, 1831], para. 224ff.).

27. E. Renan, "Claude Bernard," in Renan, ed., *L'Oeuvre de Claude Bernard* (Paris, 1881), p. 33. My italics.

28. T. H. Huxley, *Hume* (London, 1894), p. 65.

29. My labels are, of course, anachronistic. The concepts they denote are not.

30. For Peirce's application of SCT to the history of science, see his *Lessons from the History of Science,* (c. 1896), in *Collected Papers,* 1:19–49, especially para. 108, p. 44.

31. *Collected Papers,* 5:579.

32. Ibid., 5:582.

33. Ibid., 6:526; cf. also 2:755.

34. Ibid., 7:110.

35. Ibid., 2:775.

36. Ibid., 2:729.

37. Ibid., 2:769.

38. Ibid., 1:67.

39. Ibid., 5:576. Other relevant passages would include:

1868 (revised 1893): "we cannot say that the generality of inductions are true, but only that in the long run they approximate to the truth" (5:350).

1898: "a properly conducted inductive research corrects its own premises" (5:576).

1901: "[Induction] commences a proceeding which must in the long run approximate to the truth" (2:780).

"persistently applied to the problem [induction] must produce a convergence (though irregular) to the truth" (2:775).

"the method of induction must generally approximate to the truth" (6:100).

1903: "The justification of [induction] is that, although the conclusion at any stage of the investigation may be more or less erroneous, yet the further application of the same method must correct the error" (5:145).

"Suppose we define Inductive reasoning as that reasoning whose conclusion is justified . . . by its being the result of a method which if steadfastly persisted in must bring the reasoner to the truth of the matter or must cause his conclusion in its changes to converge to the truth as its limit" (7:110).

". . . if this mode of reasoning [viz., induction] leads us away from the truth, yet steadily pursued, it will lead to the truth at last" (7:111).

See also *Collected Papers,* 2:709.

40. Ibid., 2:756.

41. Ibid., 2:758.
42. Ibid., 2:77ff.
43. Ibid., 2:770.
44. Provided, of course, that there is some limit to the sequence in question; a qualification which Peirce realized to be essential.
45. Ibid., 5:574. Peirce's example, that of the extraction of roots, is identical to Hartley's and LeSage's.

It is perhaps appropriate to add here that Peirce knew Hartley's *Observations on Man* first-hand, and makes numerous references to it in his *Collected Papers*. Moreover, he knew of LeSage's work, at least second-hand, citing it in volume 5 of his *Collected Papers*.

I know too little about Peirce's intellectual biography to assert with any confidence that it was definitely Hartley and LeSage who gave him the idea of a SCM; but, given Peirce's knowledge of Hartley and the obvious similarities in the initial approaches to the problem it seems a reasonable conjecture that Hartley may have stirred Peirce to consider the question of self-correction in detail.

46. For references to the vast body of technical literature on the straight rule, cf. the bibliography in Salmon's *The Foundations of Scientific Inference* (Pittsburgh, 1967).
47. Whether that replacement is closer to the truth than that which it replaces, is, of course, another matter. But at least quantitative induction can specify a replacement, and is thus (potentially) a strongly self-corrective method.
48. This point, viz., that qualitative induction is not (or, at least, has not been shown to be) self-corrective, has gone unnoticed by several of Peirce's commentators. For instance, Cheng writes:

To say that a qualitative induction is self-correcting is either to say that a given hypothesis is replaceable by a new hypothesis or that the scope of the given hypothesis is modifiable or limitable.... (*Peirce's and Lewis's Theories,* p. 73.)

In arguing this point, Cheng has used an unfortunate sense of "self-correcting." That an hypothesis is replaceable or "modifiable" merely means that we have techniques for discarding or altering it. If qualitative induction is to be self-correcting then we need, at a minimum, the further assurance that its replacement or altered expression is an improvement. This assurance Peirce nowhere provides, and on occasion even denies that we can obtain it.

49. *Collected Papers,* 2:771.
50. Ibid., 2:759.
51. Ibid., 5:578.
52. The titles of these 32 papers are listed in Appendix I to Cheng's *Peirce's and Lewis's Theories of Induction.* Ironically, Cheng himself discusses Peirce's work as if it were designed explicitly as a reply to Hume.

53. *Collected Papers,* 7:114–19.

54. J. W. Lenz, "Induction as Self-Corrective," in E. Moore and R. Robin, eds., *Studies of Peirce,* n. 2, p. 152. Cheng echoes Lenz when he observes that "Peirce does not make clear what the self-correcting process of induction means . . ." (*Peirce's and Lewis's Theories of Induction,* p. 67).

55. One could schematically survey the major changes in SCT by looking at three formulations, the first, typically eighteenth-century, the second, nineteenth-century, and the third, Peirce's:

SCT_1: The methods of science are such that, given a refuted hypothesis H, a mechanical procedure exists for generating a "truer" H′. ∴ Science is progressive (i.e., getting closer to the truth).

SCT_2: The methods of science are such that, given a refuted hypothesis H, we can always determine whether an alternative H′ is "truer." ∴ Science is progressive.

SCT_3: The method of enumerative induction is such that, given a refuted H (and the available evidence) we can mechanically produce an alternative H′ which is likely to be "truer" than H. ∴ Science is progressive.

The sequence SCT_1–SCT_2–SCT_3 is one in which the premises become increasingly precise and defensible; but the price paid is that the premises seem to lend less and less inferential support to the conclusion.

56. *Collected Papers,* 5:582.

57. Ibid., 1:81. Peirce insists "that it is a primary hypothesis . . . that the human mind is akin to the truth in the sense that in a finite number of guesses it will light upon the correct hypothesis" (Ibid., 7:220).

58. Ibid., 6:531. Cf. also 1:121.

59. Pierre Duhem, *The Aim and Structure of Physical Theory,* trans. Wiener (New York, 1962), pp. 26–27.

60. Ibid., p. 297. Duhem summarizes his position when he observes:

To the extent that physical theory makes progress, it becomes more and more similar to a natural classification which is its ideal end. Physical theory is powerless to prove this assertion is warranted, but if it were not, the tendency which directs the development of physics would remain incomprehensible. Thus, in order to find the title to establish its legitimacy [as an SCM], physical theory has to demand it of metaphysics (Ibid., p. 298).

61. My suspicion, which I hope to investigate shortly, is that this "cheap" form of inductive self-correction has its origins in Laplace's rule of succession, and the discussions that rule engendered in nineteenth-century probability theory.

62. Reichenbach writes: "The method of scientific inquiry may be considered as a concatenation of [enumerative] inductive inference . . ." (*Experience and Prediction,* [Chicago, 1938], p. 364).

63. Cf. G. H. von Wright, *The Logical Problem of Induction,* 2nd ed. (Oxford, 1965), chap. viii.

It is, however, to von Wright's credit that he, almost alone among Peirce's commentators, perceives the limited scope of Peirce's treatment of self-correction. As he puts the point: "the Peircean idea of induction as a self-correcting approximation to the truth has no immediate significance . . . for other types of inductive reasoning than statistical generalization" (Ibid., p. 226).

64. See, for instance, W. Salmon, "Vindication of Induction," in H. Feigl and G. Maxwell, eds., *Current Issues in the Philosophy of Science* (New York, 1961), p. 256; and W. Salmon, "Inductive Inference," in B. Brody, ed., *Readings in the Philosophy of Science* (Englewood Cliffs, N.J., 1970) p. 615.

65. A very different formulation of SCT has been developed by Karl Popper in his *Conjectures and Refutations* (London, 1963). Popper's approach, unlike that of Peirce, Reichenbach and Salmon, does not attempt to make enumerative induction the cornerstone of scientific inference. It depends, rather, upon showing (unsuccessfully, I believe) that the method H-R-H' is weakly self-corrective in virtue of methodological convention about the respective domains of H and H'. Popper is perhaps alone among contemporary philosophers of science in facing the issues raised by SCT in their full generality. As inadequate as his discussion of "verisimilitude" is, he has sensed the magnitude of the problem. In this, as in other ways, Popper is probably closer to the nineteenth-century methodological tradition than is any other living philosopher.